Recalcitrant Pollutants Removal from Wastewater

Recalcitrant Pollutants Removal from Wastewater examines the role of indigenous microbes in the degradation and detoxification of wastewater utilizing the latest biological treatment technologies. It emphasizes environmental sustainability as a core theme in facilitating the adoption of circular economy objectives by industry and policymakers alike. Numerous environmentally sound strategies for industrial wastewater treatment are presented throughout, as well as practical applications for treated wastewater, including irrigation, aquaculture, and agricultural uses. Further, this book highlights best practices for the most cost-effective approaches for wastewater bioremediation technologies, as well as life-cycle evaluations of advanced wastewater detoxification approaches for restoration.

- Covers the most advanced and innovative approaches for the management of toxic compounds in industrial wastewaters.
- Describes how microbes can be helpful in successfully removing heavy metals from wastewater.
- Discusses bioremediation approaches frequently used for the mass biotechnological treatment of agricultural fields.

Recalcitrant Pollutants Removal from Wastewater

Edited by
Izharul Haq and Maulin P. Shah

CRC Press
Taylor & Francis Group
Boca Raton London New York

CRC Press is an imprint of the
Taylor & Francis Group, an **informa** business

Designed cover image: Shutterstock

First edition published 2025
by CRC Press
2385 NW Executive Center Drive, Suite 320, Boca Raton FL 33431

and by CRC Press
4 Park Square, Milton Park, Abingdon, Oxon, OX14 4RN

CRC Press is an imprint of Taylor & Francis Group, LLC

© 2025 selection and editorial matter, Izharul Haq and Maulin P. Shah; individual chapters, the contributors

ISBN: 978-1-032-43616-6 (hbk)
ISBN: 978-1-032-43617-3 (pbk)
ISBN: 978-1-003-36812-0 (ebk)

DOI: 10.1201/9781003368120

Typeset in Times
by codeMantra

Contents

About the Editors

Dr Izharul Haq is an Assistant Professor at Dr. B. Lal Institute of Biotechnology, Jaipur, India. He was previously employed as a Post-Doctoral Fellow at the Indian Institute of Technology Guwahati, India. He obtained his Ph.D. in Microbiology from CSIR-Indian Institute of Toxicology Research, Lucknow, India. He is working on liquid and solid waste management through microorganisms and their toxicity evaluation. He has 65 publications with h index of 18 and >1600 citations.

Dr Maulin P. Shah is currently working as a Deputy General Manager–Industrial Wastewater Research Lab, Division of Applied and Environmental Microbiology Lab at Enviro Technology Ltd., Ankleshwar, Gujarat, India. He received his Ph.D. (2002–2005) in Environmental Microbiology from Sardar Patel University, Vallabh Vidyanagar, Gujarat. He served as an Assistant Professor at Godhra, Gujarat University in 2001. He is a Microbial Biotechnologist with diverse research interests. A group of research scholars is working under his guidance on areas ranging from Applied Microbiology, Environmental Biotechnology, Bioremediation, and Industrial Liquid Waste Management to solid-state fermentation. His primary interest is the environment, the quality of our living resources, and the ways that bacteria can help to manage and degrade toxic waste and restore environmental health. Consequently, he is very interested in genetic adaptation processes in bacteria, the mechanisms by which they deal with toxic substances, how they react to pollution in general, and how we can apply microbial processes in a useful way (like bacterial bioreporters). One of our major interests is to study how bacteria evolve and adapt to use organic pollutants as novel growth substrates. Bacteria with new degradation capabilities are often selected in polluted environments and have accumulated small (mutations) and large genetic changes (transpositions, recombination, and horizontally transferred elements). His work has been focused on assessing the impact of industrial pollution on microbial diversity of wastewater following cultivation dependent and cultivation independent analysis. His major work involves isolation, screening, identification, and genetic engineering of high impact of microbes for the degradation of hazardous materials. He has more than 200 research publications in highly reputed national and international journals. He directs the Research Program at Enviro Technology Ltd., Ankleshwar. He has guided more than 100 postgraduate students in various disciplines of Life Science. He is an active editorial board member of more than 150 highly reputed journals in the environmental and biological sciences. He was the Founder Editor-in-Chief of the *International Journal of Environmental Bioremediation and Biodegradation* (2012–2014) and *Journal of Applied and Environmental Microbiology* (2012–2014) (Science and Education Publishing, USA). He is actively engaged as an editorial board member in 32 journals of high repute (Elsevier, Springer, Taylor & Francis, RSC, Wiley, KeAi, DeGruyter). He is also serving as a reviewer of various journals of national and international repute. He has edited more than 200 books on wastewater microbiology, environmental microbiology, bioremediation, and hazardous waste treatment.

Preface

The book titled *Recalcitrant Pollutants Removal from Wastewaters* intends to cover various aspects of wastewater treatment and management, and in-depth knowledge on materials and procedures to encourage the application of wastewater treatment technologies. Among the numerous environmental concerns, effective industrial waste management is among the important sectors that needs further focus. The existing framework for the treatment of bioremediation for both industries is not yet up to date to satisfy the social needs of the rising world. The aim of this book is to comprehensively explore environmentally sound strategies for industrial wastewater treatment, and treated wastewater would be useful for the development of parks, aquaculture, and agriculture. This book will discuss the application of wastewater bioremediation technologies in a cost-effective approach. This book will also describe the techno-economic study and life-cycle evaluation of advanced wastewater detoxification approaches for restoration.

The individual chapters will provide detailed updated information about the wastewater treatment and management including occurrence, source, characteristics, distribution, health risk, and their remedial strategies through conventional and advanced treatment processes for environmental safety.

Covering multiple facets about pollutants and their management through sustainable, advanced, and eco-friendly treatment processes, this book is meant to serve as a single term in us for students, researchers, scientists, professors, engineers, and professionals who aspire to work in the field of environmental science, environmental biotechnology, environmental microbiology, civil/environmental engineering, eco-toxicology, and other relevant areas of industrial waste management for the safety of the environment. All chapters in this book have been contributed by global experts in the field of hazardous pollutants toxicity, its characteristics, effects, and emerging detection and treatment methods.

We are grateful to the authors for compiling the pertinent information required for chapter writing, which we believe will be a valuable source for both the scientific community and the audience in general. We are thankful to the expert reviewers for providing their useful comments and scientific insights, which helped shape the chapter organization and improved the scientific discussions, and overall quality of the chapters. We sincerely thank the CRC Press team comprising the Senior Book Acquisition Editor, Editorial Project Manager, and the entire CRC Press production team for their support in publishing this book.

Contributors

Safia Ahmed
Applied and Environmental
  Microbiology Lab, Department of
  Microbiology
Quaid-i-Azam University
Islamabad, Pakistan

Qurban Ali
Applied and Environmental
  Microbiology Lab, Department of
  Microbiology
Quaid-i-Azam University
Islamabad, Pakistan

Anand Kumar J
Department of Chemical Engineering
National Institute of Technology Raipur
Raipur, India

Shreya Anand
Department of Bioengineering &
  Biotechnology
Birla Institute of Technology
Mesra, India

Anfal Arshi
Defence Institute of Bio-energy
  Research, DRDO
Haldwani, India

Suchita Atreya
Defence Institute of Bio-Energy
  Research, DRDO
Haldwani, India

Malik Badshah
Sustainable Bioenergy and Biorefinery
  Lab, Department of Microbiology
Quaid-i-Azam University
Islamabad, Pakistan

Sadhana Balaji
Department of Biotechnology
Rajalakshmi Engineering College
Chennai, India

Netra Prova Baruah
Environmental Chemistry Laboratory,
  Resource Management and
  Environment Section, Life Science
  Division
and
Institute of Advanced Study in Science
  and Technology
Guwahati, India

Arundhuti Devi
Environmental Chemistry Laboratory,
  Resource Management and
  Environment Section, Life Science
  Division
Institute of Advanced Study in Science
  and Technology
Guwahati, India

Divya Lakshmi S
Department of Biotechnology
Rajalakshmi Engineering College
Chennai, India

Boojhana Elango
Department of Microbiology
Muthayammal College of Arts and
  Science, Rasipuram
Namakkal, India

Sougata Ghosh
Department of Microbiology, School of
  Science
RK University
Rajkot, India
and
Department of Physics, Faculty of
  Science
Kasetsart University
Bangkok, Thailand

Manisha Goswami
Environmental Chemistry Laboratory,
  Resource Management and
  Environment Section, Life Science
  Division
Institute of Advanced Study in Science
  and Technology
Guwahati, India

Izharul Haq
Department of Biotechnology
Dr. B. Lal Institute of Biotechnology
Jaipur, India

Rohan Jeyakumar
Department of Chemical Engineering
KPR Institute of Engineering and
  Technology
Coimbatore, India

Dharani Loganathan
Department of Chemical Engineering
KPR Institute of Engineering and
  Technology
Coimbatore, India

Maghimaa Mathanmohun
Department of Microbiology
Muthayammal College of Arts and
  Science, Rasipuram
Namakkal, India

Jyotsana Maura
Defence Institute of Bio-energy
  Research, DRDO
Haldwani, India

Judith Infanta Madhalai Muthu
Department of Chemical Engineering
KPR Institute of Engineering and
  Technology
Coimbatore, India

Gokila Muthukrishnan
Department of Chemical Engineering
KPR Institute of Engineering and
  Technology
Coimbatore, India

Maitri Nandasana
Department of Microbiology, School of
  Science
RK University
Rajkot, India

Gunadhor Singh Okram
UGC-DAE Consortium for Scientific
  Research
University Campus, Khandwa Road
Indore, India

Padmini Padmanabhan
Department of Bioengineering &
  Biotechnology
Birla Institute of Technology
Mesra, India

Awais Qasim
Applied and Environmental
  Microbiology Lab, Department of
  Microbiology
Quaid-i-Azam University
Islamabad, Pakistan

Umapriya Rohan
Department of Chemical Engineering
KPR Institute of Engineering and
   Technology
Coimbatore, India

Biju Prava Sahariah
Department of Chemistry
National Institute of Technology Raipur
Raipur, India

Saranya Sankaran
Department of Biotechnology
Rajalakshmi Engineering College
Chennai, India

Saranya J
Department of Electronics and
   Communication Engineering
Rajalakshmi Engineering College
Chennai, India

Warda Sarwar
Applied and Environmental
   Microbiology Lab, Department of
   Microbiology
Quaid-i-Azam University
Islamabad, Pakistan

Maulin P. Shah
Industrial Wastewater Research
   Lab, Division of Applied and
   Environmental Microbiology, Enviro
   Technology Limited
Ankleshwar, India

Anshu Singh
Defence Institute of Bio-energy
   research, DRDO
Haldwani, India

Surbhi Sinha
Amity Institute of Biotechnology
Amity University Uttar Pradesh
Noida, India

Anitha Thulasisingh
Department of Biotechnology
Rajalakshmi Engineering College
Chennai, India

Aparna Yadu
Central Institute of Petrochemicals
   Engineering & Technology Raipur
Raipur, India

1 Bioengineering Approach Imbibed in Microorganisms toward Wastewater Treatment

Anitha Thulasisingh, Divya Lakshmi S, Saranya J, Saranya Sankaran, and Sadhana Balaji

1.1 INTRODUCTION

Our environment has been exposed to massive amounts of toxins such as heavy metals, dyes, hydrocarbons, and other pollutants because of anthropogenic acts due to population and industry expansion that is occurring quickly. Technology has advanced to a point where these pollutants can be removed or degraded due to the tremendous harm they pose to all living things. Due to their resilience to extreme environmental conditions, microbes play a key role in this degrading process. Microbial community includes bacteria, algae, fungi, viruses, and protozoans that interact with the elements present in the wastewater. Microorganisms are employed in wastewater treatment as they consume the nutrients present for their growth and metabolic activities or accumulate them, and thereby clear the water zone (Haq and Kalamdhad, 2021, 2023; Haq et al., 2016a, b, 2017, 2022; Haq and Raj, 2018). The advancement of genetic engineering and molecular biology led to bioengineering techniques that improved the effectiveness of microorganisms in degrading contaminants at a quicker pace, even if natural strains had the ability to degrade them. Bioengineered bacteria are mostly employed in wastewater treatment as distinct strains, and genetic modification including their role in wastewater treatment are focused in the text (Yaashikaa et al., 2022).

Along with bioengineered microorganisms, response of magnetotactic bacteria to the toxins present in the effluent is also briefed. Excellent catalytic activity, increased reactivity, and adsorptive properties are all hallmarks of nanomaterials. Therefore, by using various wastewater treatment techniques such adsorption, photocatalysis, and electrochemical treatment, nanomaterials lower the concentration of the pollutant in the water. The wastewater treatment methods can use a variety of nanomaterials, such as nanoparticles, nanorods, nanotubes, nanofibers, nanoribbons, nanosheets, quantum dots, and others. Engineered nanoparticles such as silver nanoparticles, and their antimicrobial contribution in wastewater treated, taking account of the synthesis of nanoparticles from various biological source including microorganisms and the influence of nanoparticles on microorganisms that is affected by serval factors are highlighted (Saravanan et al., 2022a).

1.2   HANDLING OF WASTEWATER

The wastewater from any source holds substances that are fatal to humans and other living organisms. The wastewater must be treated before discharging it into natural waterbodies. Industrial effluents, domestic sewage water and wastewater from hospitals, research institutions must be treated properly according to the guidelines specified by authorized institutions and government. The wastewater must be treated to removal harmful elements, contaminants, and the pathogens present in it (Haripriyan et al., 2022).

The treatment of wastewater involves several steps in a chamber provided with ports for inlet and outlet. The following procedure is worked with certain criteria and is depicted in Figure 1.1.

1.2.1   COLLECTION OF WASTEWATER

The wastewater is collected from its native source in a systematic manner through drainage management framework. The collected wastewater is stored without any leakage to the surroundings. During the water treatment process, the water is passed into the chamber.

1.2.2   ISOLATION AND CULTIVATION OF DESIRED MICROORGANISM

The microorganism that is supposed to be used in water treatment is selected based on its ability to absorb the nutrients present and conversion of them into useful biomass. The selected strain must show tolerance to the toxicity and abiotic factors present in the wastewater. The organism is genetically modified by strain improvement to increase the efficiency of the organism in clarifying the wastewater. The strain must be cultured using pure culture techniques in a sterile medium.

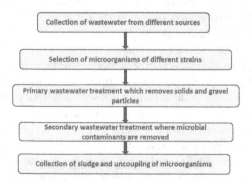

FIGURE 1.1   Wastewater treatment procedures.

1.2.3 PRIMARY TREATMENT OF WASTEWATER

Diverse techniques are used by primary wastewater treatment technologies to remove entrained and suspended particles from wastewater. To prevent damage to pumps, pipes, and other equipment downstream, heavy sediments, gravel, and sand settle in this chamber. The cleared wastewater is delivered to secondary biological treatment while the sedimented particles are collected from the tank and often routed to an anaerobic biological treatment process. To remove oil, grease, and other suspended particles, these purifiers frequently have a surface skimmer. Rapid sedimentation of particles in solid-liquid combinations is encouraged by the use of centrifugal force (Oruganti et al., 2022).

1.2.4 SECONDARY TREATMENT OF WASTEWATER

Microorganisms are used in this procedure to biologically remove pollutants from wastewater. Each process, whether aerobic, anaerobic, or even facultative, uses a unique bacterial colony. Microorganisms that have undergone genetic modification are used to react with harmful substances. They do this by precipitating or collecting the dangers from the wastewater. Nutrients like nitrogen, phosphorus, and other sources are also simultaneously absorbed (Cyprowski et al., 2018).

1.2.5 COLLECTION OF SLUDGE AND UNCOUPLING OF MICROORGANISM

The sludge is collected separately and further processed. The small amount of this activated sludge serves as inoculum for the next cycle of water treatment and the rest is used as manure. The microorganisms are separated from the treated for further level purification. Microorganisms can be separated by physical or chemical methods. The separated microorganisms are employed in other industrial processes. The clarified water is supplied to meet demands via secured pipelines.

1.3 EMPLOYMENT OF MICROORGANISMS IN WASTEWATER TREATMENT

There are various kinds of microorganisms that exist in distinct sources of nature ancestrally. Bacteria, fungi, microalgae, protozoans, virus, and yeast are diverse forms of microorganisms. They have many ecological importance in ecosystem balance. One such significance is its role in wastewater treatment. Microorganisms are capable of absorbing nutrients present in wastewater and converting it into biomass. A few microorganisms that play a significant role in wastewater treatment are discussed here (Chang et al., 2014).

1.3.1 PSEUDOMONAS AERUGINOSA

Infections caused by *Pseudomonas aeruginosa*, an important opportunistic Gram-negative human pathogen, include sepsis, burns and wound infections, ventilator-associated pneumonia, and persistent lung infections in cystic fibrosis patients.

One of the ESKAPE bacteria that is capable of rapidly acquiring multidrug resistance (MDR) is *P. aeruginosa*, which has inherent resistance to a variety of antibiotics, including certain beta-lactams.

As a result, highly resistant pathogens are now found in large quantities in water and wastewater sources. For example, *P. aeruginosa* was more prevalent than in other nations in the aquatic sediments of India receiving hospital wastewater, while carbapenem-resistant *Escherichia coli* and *P. aeruginosa* were prevalent in residential wastewater treatment facilities. Rivers frequented by people from all over India and the world are considered hotspots for gene transfer events that lead to antibiotic resistance (Sun et al., 2022).

Investigations were of *P. aeruginosa* outbreaks in wastewater from households and hospitals, and in water bodies with varying levels of wastewater contamination. Extracellular polysaccharides and intracellular polyhydroxy butyric acid produced by *P. aeruginosa* in response to osmotic stress cause coagulation of wastewater, resulting in a decrease in color. In addition, enzymes such as catalase, glucose oxidase, peroxidase, and laccase present in certain bacteria have been shown to be involved in the discoloration of wastewater (Menon et al., 2021).

1.3.2 RHODOPSEUDOMONAS SPECIES

For the first time, heavy refinery effluent was treated to remove organics and nitrogen using a mixed culture system dominated by *Rhodopseudomonas* and *Pseudomonas*. Coenzyme Q10 and other high-value impurities like pigment build up in the cell at the same time. Moreover, microbial alterations at various light intensities were examined using q-PCR and microbial community analysis. This study sought to determine if *Rhodopseudomonas* and *Pseudomonas* could effectively handle the refractory heavy refinery effluent and increase the generation of high-value contaminants in the cells. Investigated were the effects of *Rhodopseudomonas capsulata* on the biodegradation of carbaryl in the soil and the enhancement of soil fertility. With the addition of wastewater containing *R. capsulata*, carbaryl was successfully eliminated in comparison to the control treatment, and soil fertility was restored (Wu et al., 2019).

The development of *R. capsulata* could not, however, be sustained for more than a single day due to the absence of organic matter in the soil and the control treatment. *R. capsulata* in the wastewater has access to sufficient carbon and energy sources thanks to the remaining organic materials. Using this approach, carbaryl pollution is addressed, soil fertility is increased, and wastewater from the processing of soybeans is reused as sludge. A high-order nonlinear mathematical model of the rates of carbaryl elimination has been developed in the interim (Sun et al., 2022).

1.3.3 CHLORELLA VULGARIS

Algae are naturally autotrophic organisms that can produce biomass from inorganic carbon sources like carbonates and CO_2 and release oxygen into the environment. Recent years have seen an increase in the amount of research examining the use of algae, bacteria, or fungus in the treatment of wastewater, drawing attention to the application of algal biotechnology in wastewater management. Moreover, it has been shown that algae, together with microalgae and cyanobacteria, can improve aeration, aiding in the improvement of organic pollutant decomposition efficiency. is the presence of pollutants because it reduces the demand for external nutrition supply for algae. Moreover, microalgae are appropriate for bioremediation applications because of their quick growth and ability to adapt to any challenging environment.

Microalgae are becoming an essential organism in today's economy-driven world because to value-added goods they produce like ethanol, methane, and other substances utilized as organic fertilizers and animal feed. *Chlorella vulgaris* was selected as a model species due to its quick development, effective absorption of nutrients and organic matter from wastewaters, and widespread presence in wastewaters all over the world, even those in temperate temperatures. am. In the proposed integrated biphasic autotrophic mixotrophic culture method, *C. vulgaris* successfully treated concentrated effluent and produced biofuel, optimizing lipid output and attaining high nutrient and COD removal efficiencies (Ge et al., 2018).

1.3.4 RAPHIDOCELIS SUBCAPITATA

Effective treatment to effectively remove contaminants and biotoxicity is one of the major challenges in wastewater management. The test organisms from the algae *Raphidocelis subcapitata* used in the biotoxicity tests were provided by the Research Group for Aquatic Ecosystems (GEEA), Department of Environmental Engineering, Federal University of Sergipe (Brazil) and recommended. Algae are one of the major components of aquatic ecosystems, as producers of oxygen, self-cleaning in rivers and lakes, and as the main food source for important consuming organisms. The green alga *R. subcapitata* is widely distributed in freshwater ecosystems and is widely used as a model for phytoplankton (Fan et al., 2018).

1.3.5 PICHIA PASTORIS

Pichia pastoris underwent transformation and expressed recombinant Hfphytase. *P. pastoris* was heterologously expressed with lignin-degrading enzymes from several fungal sources to increase enzyme output. For instance, one study discovered that *P. pastoris* expressed the Lac gene from *Trametestrogii* and that a thermostable recombinant Lac with a half-life of 45 minutes at 70°C was able to decolorize azo pigments like Acid Red 26. Another study discovered that *P. pastoris* expressed the Lac gene from Trametes sp. 48424. Recombinant Lac was produced in high quantities and has the ability to decolorize a variety of colors. TS-A CGMCC 12964 efficiently removes azo pigments, and *P. pastoris* GS115 secretively expressed the recombinant enzyme. Recombinant Lac was produced in high quantities and has the ability to decolorize a variety of colors. TS-A CGMCC 12964 efficiently removes azo pigments, and *P. pastoris* GS115 secretively expressed the recombinant enzyme (Gao et al., 2022).

1.4 BIOENGINEERED MICROORGANISMS IN WASTEWATER TREATMENT

Microorganisms are genetically modified using genetic engineering tools like rDNA Technology to increase efficiency, tolerance to toxins and genetic stability. The genetically modified microorganism removes the hazards in wastewater with higher efficiency than that of an unmodified organism (Tang et al., 2022). Exceptionally, bioengineered microorganisms are used to remove harmful elements like heavy metals, free-radical, and so on. Now, let us discuss why microorganisms are bioengineered and how they are used in the treatment of wastewater.

1.4.1 ROLE OF BIOENGINEERED BACTERIA IN WASTEWATER TREATMENT

Most commonly the microorganisms used in wastewater treatment are bacteria, especially it is anaerobic bacteria. Anaerobic bacteria are used because they handle the fermentation of methane of sewage sludge which eases the decomposition of macromolecules of organic compounds into the simpler compounds.

Bioengineering is a process of modifying genetic material to obtain a new species which may be useful in doubling up the process. In the case of wastewater treatment bioengineered bacteria is used which may help to reduce pathogenic bacteria and dangerous chemicals and pollutants which results in a clean water source for human, agriculture and for industrial use. These bioengineered bacteria are not only used in wastewater treatment but also to produce enzymes, flavoring agents, and other compounds. In general, bioengineered microbes have greater shelf life, cost effective, produce higher yields, and used in medicines for producing life-saving vaccines, insulin, and treatment for diseases.

The bioengineered bacteria are used in wastewater treatment as the genetic material of the organism is modified to produce certain enzymes and protein that removes the hazardous elements from the water. Mostly the treatment of industrial effluents employs bioengineered bacteria to eliminate the virulent elements (Malik et al., 2022; Deans, 2014).

1.4.2 BACTERIAL STRAINS

There are various species of bacteria present all around the world ancestrally. The bacteria can be seen in marine, freshwater, aquatic, and terrestrial conditions. The bacteria can absorb the nutrients present in wastewater and convert it into useful biomass. The bacteria employed in wastewater treatment can be aerobic, anaerobic, or facultative in nature. The bacterial species mostly associated with water treatment are as follows:

- *Escherichia coli*
- *Rhodopseudomonas*
- *Pseudomonas aeruginosa*
- *Methanobacterium*
- *Methanobrevibacterium*
- *Clostridium perfringens*
- *Enterobacter cloacae*
- *Enterococcus faecalis*
- *Klebsiella pneumoniae*
- *Proteus vulgaris*

The bioengineered *Escherichia coli* is mostly used in wastewater treatment. The contributions of bioengineered bacteria in wastewater treatment, especially *E. coli* and other important bacteria used in wastewater treatment are discussed below (Li et al., 2017).

1.4.3 ESCHERICHIA COLI

According to recent research, a few strains of *E. coli* bacteria are not only surviving, but also thriving in wastewater treatment plants. The K-12 *E. coli* strain, which manages most severe infections but lacks the surface features required to effectively colonize the human gut, is a frequently researched laboratory organism. Moreover, the gene that codes for the toxin that causes it is absent. Even though the K-12 strain is thought to be non-pathogenic, it must be treated with the right microbiological procedures and safety measures. *E. coli* is a suggested host for gene cloning due to the efficiency with which DNA molecules are introduced into cells. Mostly, the *E. coli* K-12 strain is used to clean wastewater. To develop a specific protein that controls the manufacture of desired enzymes, *E. coli* was genetically altered. The enzyme generated is capable of removing heavy metals from wastewater. Heavy metals like cadmium and lead can be removed from effluents by precipitation or accumulation using genetically engineered *E. coli*.

Although chlorination, oxygenation, and other treatments used in sewage treatment facilities kill *E. coli*, the researchers discovered that some strains of dangerous *E. coli* did not. It was chosen as the experimental organism because the bacteria grow quickly in growth conditions with certain chemical compositions and because the cells do not congregate. Due to its large genetic repertoire and capacity to thrive on chemically specified medium, *E. coli* has become a crucial system in the investigation of bacterial metabolic pathways (Rutherford, 2021).

1.4.4 STRAIN IMPROVEMENT TECHNIQUES

Using recombinant DNA technology, strain improvement is a sophisticated biotechnological process that increases the yields of healthy metabolites for people. By enhanced substrate usage, altered enzyme activity, and increased resistance to phage infection, strain enhancement seeks to increase product quality and production. Among the principal genetic routes for strain improvement are:

1. mutation to produce a new genetic variant
2. application of screening techniques to select desired improved strains
3. improved strains are identified
4. optimization and downstream processing of operated mass cultures and cellular responses (Fiedurek et al., 2017)

1.4.5 GENETIC MODIFICATION OF BACTERIA

The bacteria that are bioengineered are accomplishable for removal of toxic substance present in the wastewater. Natural bacterial species can be modified genetically by strain improvement techniques to improve their efficiency to high-yielding microorganisms, i.e., mutants that exhibit resistance to anti-metabolites, nutritional inhibitors, and abiotic stresses as well as tolerance to solvents and substrates or products toxic. Not all species can undergo genetic modification. The bacterial species must fulfill the following criteria:

1. must have high genetic stability.
2. must be capable of degrading organic elements within a short span.
3. should be cultivated easily in a sterile medium (Konar and Datta, 2022).

Recombinant DNA (rDNA) technology, which includes changing genetic material outside of an organism to produce desired and better features in live creatures or as their products, is used to genetically modify the bacterium. Using this approach, DNA fragments with the required gene sequence are inserted through an appropriate vector from a range of sources (Khan et al., 2016).

The selection of a desirable bacterial strain is the initial stage in the rDNA technology process. Let us operate bacteria *E. coli* using rDNA technology to obtain desired property. The gene of interest is isolated from another organism and inserted into the bacterial plasmid DNA. Cut the gene of interest and plasmid DNA using same restriction endonuclease enzyme such that the gene of interest fits into the plasmid DNA like that of lock and key. Then, the enzyme DNA ligase is added to join the DNA pieces by forming base pairs. Thus, the recombinant DNA plasmid is obtained and inserted into the *E. coli*. Certain genetic properties of *E. coli* are altered. Finally, the transformed organism is cultured in a sterile medium to increase the bacterial population of that strain (Saravanan et al., 2022b).

1.4.6 APPLICATION OF MAGNETOTACTIC BACTERIA IN WATER TREATMENT

Diverse Gram-negative mobile aquatic microorganisms known as magnetotactic bacteria have the ability to physiologically multimineralize chromosomes, which are nanoscale internal magnetic crystals. In a magnetic field, known as a magnetic field, cells passively align and swim along magnetic field lines, causing them to accumulate in the edge of the droplet in the magnetic field when observed under a microscope. These organisms were discovered based on their magnetic response to the magnetic field (Wang et al., 2020).

Aquaspirillum (now Magnetospirillum) magnetotacticum MS-1 strain, a magnetically generated spirulina, was the first magnetotactic bacteria (MTB) to be cultivated in axenic media. Exogenous materials, such plasmid structures, may be more effectively introduced into organisms under user-defined circumstances using nano- and microtechnology. One illustration is the penetration of nanochannels, microchips, and electrophoresis. Instead of using *E. coli* conjugations or other biological techniques, penetration and electrophoresis employ physical forces to temporarily create pores in the cell wall or membrane or locally penetrate the cell. High-speed centrifugation of bacteria allows them to mechanically enter nano- and microstructures. Local potential changes between cells are created by electroporation using sudden voltage variations. This causes transient holes to emerge in cell walls and membranes, allowing biomolecules like DNA to enter and undergo transformation (Tay et al., 2019).

In particular for MTB strains that are challenging to transform, this method can change MTB with greater precision in DNA/bacterial copy number than random mating with *E. coli*. For instance, the use of bulk liquid electrophoresis to insert DNA plasmids into the MTB; it is thought that this procedure may be enhanced by employing microfluidic electrophoresis.

For many heavy metals in wastewater, MTB has a high biosorption capability. Several studies looked at the recycling of heavy metals using MTB after MTB was discovered. Except for Hg^{2+}, various MTB strains were chosen for the treatment of wastewater containing heavy metals with removal rates ranging from 38.6% to 100.0%. For Au^{3+}, Cr^{6+}, and Cu^{2+}, most of the published publications on heavy metal removal by MTB concentrate on specific heavy metals. Just a small number of studies have examined the capacity to remove heavy metals from wastewater. To remove Au^{3+} from polluted wastewater, MTB is utilized. Stenotrophomonas species, a MTB, was identified by Songs and his colleagues from a municipal wastewater treatment facility and utilized to concentrate Au^{3+} in demineralized water (Vargas et al., 2018).

1.4.7 Impact of Microorganisms in Water Treatment

Native bacteria are less able or unable to remove toxic compounds. Efforts have been made to increase the pollutant removal capacity of these native bacteria through the interdisciplinary involvement of molecular, genomic, bioinformatics, and microbiological techniques by editing genetically modified organisms are called genetically modified organisms. The success of the use of genetically modified organisms depends on their survival under ecological stress and, when the desired goal has been achieved, proper mechanisms for their elimination must be in place. it out of the impact zone (Torres et al., 2019).

The use of microorganisms as heavy metal bio-sorbents is an alternative to the chemical methods. However, conventional wastewater treatment processes lack specificity in binding metals, which can lead to difficulties in recovering and recycling the desired metal(s). Developments in genetic engineering supply new means of cleaning metals. Employment of genetic modification in microorganisms can improve not only their specificity but also their ability to accumulate heavy metals.

The quantities of organic and inorganic species that can compete in the biosorption process, the amount of carbon supply, and the affinity for metal ions all have a significant impact on the removal of heavy metals from wastewater using microorganisms. To improve the sorption capacity of different metals and their removal method, recombinant bacteria have recently been studied for application in heavy metal removal (Liwarska-Bizukojc, 2022).

Introduction of a single gene can alter the expression of a gene, as introduction of genes into the genome of other organisms yields different results. Gene sequences and their functions in the donor organism play a well-organized role in the microorganism from which it was isolated. Gene copies can undergo mutations such as insertions, integrations, and deletions, resulting in loss of specific functions such as instability and disruption of other gene functions (Haripriyan et al., 2022).

1.4.8 SIGNIFICANCE OF UTILIZING BIOENGINEERED MICROORGANISMS

Microorganisms have a great deal of promise for cleaning up polluted wastewater. In addition, bioprinting is paired with potent genetic engineering tools, and such a method can produce tailored microbial systems with needed or improved capacities to eliminate harmful pollutants. The bioengineering method overcomes biological water treatment's limitations and creates new opportunities for the economical, effective, and environmentally responsible treatment of wastewater. After treatment, the wastewater is free of germs and hazardous substances. This makes groundwater recharge better.

Due to their cheap capital, energy, and material costs, the employment of bioengineered microbes in the remediation of polluted water frequently has a positive economic impact. By utilizing bioengineered microorganisms to clean wastewater, it is possible to raise groundwater recharge rates and improve water quality. Byproducts of the water treatment process produced by bioengineered microbes include biomass or biofuel. The sludge is utilized as a biofertilizer in crop areas in addition to serving as an inoculum for the subsequent cycle of water treatment. (Ravikumar et al., 2021; Gautam et al., 2019)

1.4.9 ENGINEERED ANTIMICROBIAL NANOPARTICLES AND THEIR APPLICATIONS IN WASTEWATER TREATMENT

Catheters commonly cause bloodstream infections in clinics, which can result in severe illnesses. An environmentally friendly technique was developed and combined with *Staphylococcus aureus* to coat central venous catheters with silver nanoparticles (AgNPs). The diameter of the bacterium inhibition zone ranged from 0.2 to 0.35 cm (about 0.14 in), and it demonstrated improved efficacy against bacteria. Silver-impregnated catheters showed significantly decreased infection rates when compared to non-impregnated catheters (Lajcak et al., 2013). As a result of their interaction with experimentally defined Ag-chitosan nanoparticle coated metal dental implants in External Ventricular Drains, two significant dental infections, *S. mutans* and *P. gingivalis*, were successfully suppressed. Observations include reduced biofilm development, decreased toxicity, and suppression of adherence to bacteria (Divakar et al., 2018).

Titanium oxide added to concrete mix inhibits fungal colonization and fouling on concrete walls and structures when exposed to sunlight. With long-lasting antibacterial activity, graphene oxide (GO) and reduced graphene oxide (rGO) cause *E. coli* to lose 90% of its viability, which in turn causes physical damage to bacterial cells with reduced resistance and toxic effects. Antibacterial nanoparticle production is thought to be challenging on the bacterial cell. Van der Waals forces, hydrophobicity, ligand-receptor binding, and electrostatic interaction are the main factors in nano-bacterial biointerface (Wang et al., 2017). Mice were given a single dosage of 10 mg/kg of silver nanoparticles, which were 10, 40, and 100 nm in size and had citrate and polyvinyl pyrrolidine (PVP) carrying negative charge on their surfaces. Midzonal hepatic cellular necrosis and size-dependent toxicity are among the results (Recordati et al., 2016). There is apoptosis in the lungs and kidneys as a result of endoplasmic reticulum stress after 2 g of silver nanoparticles with a diameter of 20 nm are exposed to a mouse cell line in an in vitro setting (Huo et al., 2015).

During intratracheal inhalation by an adult and young rat with substantial lung damage, gold nanoparticles coated with PVP measuring 5,100 nm were detected (Tsuda et al., 2019). Nanomaterials are characterized by the production of singlet oxygen, the considerable absorption of near-infrared radiation, and the acceptable conversion of light to heat (NMs). With regard to MDR (multidrug-resistant) pathogens, Methicillin-resistant S. Aureus, different groups of MDR (multidrug-resistant) Gram-positive (MRSA, MRSE, and MLSB), and Gram-negative [extended-spectrum betalactamase (ESBL)], AmpC, and CR pathogens, Au NPs has demonstrated superior antimicrobial activity (Malarkodi et al., 2014). Excellent antibacterial action against *E. coli* 0157:H7, *S. aureus*, *L. monocytogenes*, *S. enteritidis*, and *P. fluorescens* has been demonstrated by TiO_2 NPs. Remarkable antifungal activity has been demonstrated against Candida spp., *P. expansum*, *A. niger* spp., and *P. oxalicum* (Rajeswari et al., 2021).

The engineering and scientific communities throughout the world have seen significant advancements in recent years due to the nanoscience and technology. Silver nanoparticles are mostly used for environmental hazards prevention and human health protection. Silver nanoparticles are highly helpful in the treatment of environmental and industrial effluent wastewater due to their efficiency in disinfecting the air, surfaces, and water (Chmielewska et al., 2021). The use of zinc oxide nanoparticles as an antibacterial agent in medicine shows impressive advancements in their ability to guard against environmental dangers. The discharge of ZnO nanoparticles into aquatic environments through home and industrial wastewaters has the potential to cause harmful impacts on fish and other creatures.

Bacillus subtilis, *Pseudomonas fluorescens*, and *E. coli* have all been used as test subjects for the antibacterial effectiveness of ZnO particles (Anjali Das et al., 2020). Iron oxide because of the nanoflower petals have a higher surface area-to-volume ratio and hence more active sites for analyte molecule adsorption, it is more sensitive than other zinc oxide nanoparticles and thin films.

The ozonation process has been utilized extensively in wastewater treatments such as the breakdown of hazardous organic contaminants and disinfection. By oxidizing contaminants found in water and wastewater and creating hydroxyl radicals, ZnO catalysts are widely used in the catalytic ozonation process (Amna, 2018).

Methicillin-resistant *S. aureus* and *Bacillus subtilis* were used to investigate the antibacterial effectiveness of silver nanoparticles with sizes between 4 and 15 nm. *Bacillus subtilis*' Minimum Inhibition Concentration (MIC) was 40 g/mL and methicillin-resistant *S. aureus*' MIC was 55 g/mL, respectively, at a 24-h incubation period (Das et al., 2013). The antibacterial activity of Ag NPs with diameters ranging from 1 to 100 nm is investigated. The research revealed that only Ag NPs with a 1–10 nm size range were hazardous. In addition to the size and shape of Ag NPs, a research by Anuj et al. (2020) has emphasized the effect of bacterial size on the antibacterial action of Ag NPs (Anuj et al., 2020).

P. aeruginosa microbe acquired excellent antibacterial activity with Ag NPs with particle sizes range from 5 to 10 nm for 1–5 g/mL MIC (Liao et al., 2019). At an incubation time of 24 hours, iron nanoparticles measuring 120 nm demonstrated remarkable antibacterial activity against *S. aureus* with MICs of 1–5 and 1–2 g/mL against *P. aeruginosa* (Lustosa et al., 2017). Fe-Ag nanoparticles with a core-shell structure and sizes between 10 and 30 nm demonstrated advanced antimicrobial activity with MICs between 100 and 540 mg/L against *Enterococcus faecalis*, *S. aureus*, *E. coli*, *P. aeruginosa*, *S. epidermidis*, *Enterococcus faecium*, and *Kle* Given that the distinctive bimetallic Fe-Ag NPs can be readily magnetically separated, they represent a very promising material for cutting-edge antibacterial and reductive water treatment technologies. A larger range of fungal infections, including *A. niger*, *A. flavus*, *A. fumigatus*, *A. terreus*, *P. chrysogenum*, *F. solani*, and *L. theobromae*, were investigated for the fungicidal activity of ZnO NPs. *A. terreus* had the least amount of inhibition, whereas Theobromae had the most. It was also discovered that the antifungal activity increased linearly with NP concentration and that the inhibition was dose-dependent. This happens as a result of the increased ROS generation that the increase in NP concentration causes (Jaffri et al., 2018). The Gram-positive bacteria Bacillus subtilis was significantly more efficiently eliminated by UV-induced TiO_2 NPs than by the UV disinfection procedure alone [the UV disinfection procedure is now being studied as a potential replacement for the conventional chlorine disinfection (Saravanan et al., 2022a)].

CuO NPs have been discovered to exhibit an antibacterial effect that is stronger against bacteria than fungus, similar to many other metal and metal oxide NPs. Due to their tiny size, CuO NPs may easily diffuse through a microbial cell membrane's pores. Ion channels, endocytosis, and transporter proteins are additional routes for CuO NPs to enter a microbial cell. Once within, CuO NPs interact with oxidative organelles, producing Cu^{2+} ions and ROS. The cellular toxicity of CuO NPs is caused by these Cu^{2+} ions and ROS. By combining conventional nanoparticles (NPs) with carbon nanoforms, their antibacterial capabilities can be increased. Among them, carbon nanotubes (CNTs) and graphene (G) with its derivatives (i.e., GO and rGO) are frequently used for water and wastewater treatment because they have (i) an improved ability to adsorb a variety of contaminants, (ii) have a large specific surface area, (iii) have quick degradation kinetics, and (iv) have quick charge transfer and suppressed hole-electron recombination under irradiation (Lu et al., 2016).

1.5 WASTEWATER TREATMENT THROUGH pH AND TEMPERATURE ADJUSTMENT

Heavy metal and hazardous organic compound removal from effluents is a standard part of wastewater treatment. Here, by adding acidic or basic chemicals, submerged garbage may be separated from water as part of treating the wastewater. In general, positive hydrogen ions and negative hydroxide ions make up water. Positive hydrogen ions will be more concentrated in an acidic medium (pH 7), neutral medium (pH = 7) will have an equal concentration of both positive hydrogen ions and negative hydroxide ions, and basic media (pH > 7) will have an increased concentration of negative hydroxide ions.

Negatively charged hydroxide will form bonds with positively charged metal oxide by increasing the pH of wastewater (contaminants). This makes it easier to eliminate impure particles that settle to the bottom using the proper filtration methods.

Water pollutants will float and not settle at an acidic pH, but too many positive hydrogen ions will kill microorganisms. At a pH of 7, water molecules will form, and the pollutants will stay. At basic pH levels, water pollutants can be successfully removed by filters; yet a modest basic pH favors the destruction of living cells that come into contact. So, following wastewater treatment, it is crucial to keep the water in a neutral medium (Gómez-Ortíz et al., 2013).

The effect of pH on peroxidase activity was investigated across a broad pH range (1.5–12). The results confirmed that free peroxide decreased significantly below and above pH 6 and that enzyme activity peaked at this pH. On the other hand, alkaline environments increase the stability of magnetic nanoparticles with immobilized peroxide. This is mostly caused by the pH's interaction with the enzyme, which affects functional groups by ionizing them (Darwesh et al., 2019). The pH of the solution is one element that is crucial for the photocatalytic degradation of azo dyes. The properties of ZnO's surface charge, the charge of azo dye molecules, the adsorption of dyes onto ZnO surfaces, the size of formed aggregates, the conduction and valence band edge potentials of ZnO, and the quantity of hydroxyl radicals in solution are all impacted by it. Methyl orange, Congo red, and straight blue removal efficiencies were determined in one research utilizing a 30-ppm dye concentration and 0.2 g/L ZnO catalyst load during the first 10 minutes of irradiation time. Findings revealed that the greatest value occurred at a pH of 2. First off, because all three azo dyes are anionic, ZnO can better absorb them. Second, dye molecules are more likely to exist in their quinonoid forms than in their azo forms, which are more unstable and prone to degradation, in very acidic environments. The primary cause of getting better effectiveness at an acidic pH is due to this (Weldegebrieal, 2020).

Magnesium oxide nanoparticles (MgO-NPs) were created by *Rhizopus oryzae* utilizing an improved procedure at pH 8 and 35°C. With inhibition zones of 10.6 0.4, 11.5 0.5, 13.7 0.5, 14.3 0.7, and 14.7 0.6 mm (about 0.02 in), the developed MgO-NPs at a specific concentration of 200 g/mL were found to be an excellent antimicrobial agent against the pathogens including *S. aureus*, *Bacillus subtilis*, and *Pseudomonas aeruginosa*. In order to effectively treat wastewater and kill bacteria while removing undesired chemicals, pH and temperature are crucial factors (Hassan et al., 2021).

1.6   SUMMARY AND CONCLUSION

This chapter elucidates the contribution of bioengineered microorganisms in the treatment of wastewater from various sources. The handling procedure and effect of the water treatment varies in consideration with the usual biological treatment and the employment of genetically modified microorganisms. The gene-level modification of the organism enhances its performance and toxicity tolerance during the purification process, which is distinct. The efficiency of wastewater treatment depends on the microorganism and the strain selected. The bacteria are mostly preferred for genetic modification and effective treatment of wastewater. The execution of strain improvement techniques assists the genetic modification and gene expression of the microorganism. The MTB contribute to the treatment of wastewater in an effective way. The engineered antimicrobial nanoparticles can be employed for absorbance of contaminants and toxins. The parameters such as pH and temperature of wastewater discharged even influence the rate of biological activity. The wastewater treatment through biological activity of genetically modified microorganisms has a significant impact on ecological balance and resource management.

The employment of microorganisms in wastewater treatment is a biological and an environmentally friendly approach. The bioengineered microorganism has high tolerance to external environment and to toxic contents in the wastewater with stable genetic features. The bioengineered microorganisms absorb or accumulate nutrients and clear the wastewater with high efficiency comparatively. The advancement of genetic engineering tools and synthetic biology has led to improvised sustainable development and ecological conservation. The microorganisms based on the nature and surviving ability are found in various natural sources have been identified, isolated, and genetically altered to treat various classes of wastewater.

REFERENCES

T. Amna, Shape-controlled synthesis of three-dimensional zinc oxide nanoflowers for disinfection of food pathogens. *Z. Naturforsch. C. J. Biosci.*, 73, no. 7–8 (July 2018), p. 297.

C.G. Anjali Das, V. Ganesh Kumar, T. Stalin Dhas, V. Karthick, K. Govindaraju, J. Mary Joselin, J. Baalamurugan, Antibacterial activity of silver nanoparticles (biosynthesis): A short review on recent advances. *Biocatal. Agric. Biotechnol.*, 27 (Aug 2020), p. 101593.

S.A. Anuj, H.P. Gajera, D.G. Hirpara, B.A. Golakiya, The impact of bacterial size on their survival in the presence of cationic particles of nano-silver. *J. Trace Elem. Med. Biol.*, 61 (2020), p. 126517. https://doi.org/10.1016/j.jtemb.2020.126517.

S. Chang, H. Shu, The construction of an engineered bacterium to remove cadmium from wastewater. *Water Sci. Technol.*, 70, no. 12 (2014), p. 2015.

S.J. Chmielewska, K. Skłodowski, J. Depciuch, P. Deptula, E. Piktel, K. Fiedoruk, P. Kot, P. Paprocka, K. Fortunka, T. Wollny, P. Wolak, M. Parlinska-Wojtan, P. B. Savage, R. Bucki, Bactericidal properties of rod-, peanut-, and star-shaped gold nanoparticles coated with ceragenin CSA-131 against multidrug-resistant bacterial strains. *Pharmaceutics*, 13, no. 3, (Jan 2021), p. 425.

M. Cyprowski, A. Stobnicka-Kupiec, A. Lawniczek-Walczyk, A. Bakal-Kijek, M. Golofit-Szymczak, R. L. Gorny, Anaerobic bacteria in wastewater treatment plant. *Int. Arch. Occup. Environ. Health*, 91, no. 5 (2018), p. 571.

O.M. Darwesh, I.A. Matter, M.F. Eida, Development of peroxidase enzyme immobilized magnetic nanoparticles for bioremediation of textile wastewater dye. *J. Environ. Chem. Eng.*, 7, no. 1 (2019), p. 102805. https://doi.org/10.1016/j.jece.2018.11.049.

R. Das, M. Saha, S.A. Hussain, S.S. Nath, Silver nanoparticles and their antimicrobial activity on a few bacteria. *Bionanoscience*, 3, (2013), pp. 67–72. https://doi.org/10.1007/s12668-012-0070-5.

T. Deans, Parallel networks: Synthetic biology and artificial intelligence. *ACM J. Technol. Comput. Syst.*, 11, no. 3 (Dec 2014), p. 1.

D.D. Divakar, N.T. Jastaniyah, H.G. Altamimi, Y.O. Alnakhli, Muzaheed, A.A. Alkheraif, S. Haleem, Enhanced antimicrobial activity of naturally derived bioactive molecule chitosan conjugated silver nanoparticle against dental implant pathogens. *Int. J. Biol. Macromol.*, 108, (March 2018), p. 790.

L. Fan, M.T. Brett, B. Li, M. Song, The bioavailability of different dissolved organic nitrogen compounds for the freshwater algae Raphidocelissubcapitata. *Sci. Total Environ.*, 618 (March 2018), p. 479.

J. Fiedurek, M. Trytek, J. Szczodrak, Strain improvement of industrially important microorganisms based on resistanceto toxic metabolites and abiotic stress. *J. Basic Microbiol.*, 57, no. 6 (June 2017), p. 445.

M. Gao, Y. Zhou, J. Yan, L. Zhu, Z. Li, X. Hu, X. Zhan, Efficient precious metal Rh (III) adsorption by waste *P. pastoris* and *P. pastoris* surface display from high-density culture. *J. Hazard. Mater.*, 427 (April 2022), p. 121840.

B. Gautam, A. Rajbhanshi, R. Adhikari, Bacterial load reduction in Guheswori Sewage Treatment Plant, Kathmandu, Nepal. *J. College Med. Sci. Nepal*, 15, no. 1 (March 2019), p. 40.

S. Ge, S. Qiu, D. Tremblay, K.V.P. Champagne, P.G. Jessop, Centrate wastewater treatment with *Chlorella vulgaris*: Simultaneous enhancement of nutrient removal, biomass and lipid production. *Chem. Eng. J.*, 342 (June 2018), p. 310.

N. Gomez-Ortiz, S. De la Rosa-Garcia, W. Gonzalez-Gomez, M. Soria-Castro, P. Quintana, G. Oskam, B. Ortega-Morales, Antifungal coatings based on Ca $(OH)_2$ mixed with ZnO/TiO_2 nanomaterials for protection of limestone monuments. *ACS Appl. Mater. Interfaces*, 5, no. 5 (March 2013), p. 1556.

I. Haq, A.S. Kalamdhad, Phytotoxicity and cyto-genotoxicity evaluation of organic and inorganic pollutants containing petroleum refinery wastewater using plant bioassay. *Environ. Technol. Innov.*, 23 (2021), Article 101651.

I. Haq, A.S. Kalmdhad, Enhanced biodegradation of toxic pollutants from paper industry wastewater using Pseudomonas sp. immobilized in composite biocarriers and its toxicity evaluation. *Bioresour. Technol. Rep.*, 24 (2023), p. 101674.

I. Haq, A.S. Kalamdhad, A. Pandey, Genotoxicity evaluation of paper industry wastewater prior and post-treatment with laccase producing Pseudomonas putida MTCC 7525. *J. Clean. Prod.*, 342 (2022), Article 130981.

I. Haq, A. Raj, Markandeya Biodegradation of Azure-B dye by Serratia liquefaciens and its validation by phytotoxicity, genotoxicity and cytotoxicity studies. *Chemosphere*, 196 (2018), pp. 58–68.

I. Haq, S. Kumar, V. Kumari, S.K. Singh, A. Raj, Evaluation of bioremediation potentiality of ligninolytic *Serratia liquefaciens* for detoxification of pulp and paper mill effluent. *J. Hazard Mater.*, 305 (2016a), pp. 190–199.

I. Haq, S. Kumar, A. Raj, M. Lohani, G.N.V. Satyanarayana, Genotoxicity assessment of pulp and paper mill effluent before and after bacterial degradation using Allium cepa test. *Chemosphere*, 169 (2017), pp. 642–650.

I. Haq, V. Kumari, S. Kumar, A. Raj, M. Lohani, R. N. Bhargava Evaluation of the phytotoxic and genotoxic potential of pulp and paper mill effluent Using Vigna radiata and Allium cepa. *Adv. Biol.*, 2016 (2016b), pp. 1–10, Article ID 8065736.

U. Haripriyan, K.P. Gopinath, J. Arun, M. Govarthanan, Bioremediation of organic pollutants: A mini review on current and critical strategies for wastewater treatment. *Arch. Microbiol.*, 204, no. 286 (2022). https://doi.org/10.1007/s00203-022-02907-9.

S. E.-D. Hassan, A. Fouda, E. Saied, M.M.S. Farag, A.M. Eid, M.G. Barghoth, M.A. Awad, M.F. Hamza, M.F. Awad, Rhizopusoryzae-mediated green synthesis of magnesium oxide nanoparticles (MgO-NPs): A promising tool for antimicrobial, mosquitocidal action, and tanning effluent treatment. *J. Fungi.*, 7, no. 5 (2021), p. 372. https://doi.org/10.3390/jof7050372.

L. Huo, R. Chen, L. Zhao, X. Shi, R. Bai, D. Long, F. Chen, Y. Zhao, Y.-Z. Chang, C. Chen, Silver nanoparticles activate endoplasmic reticulum stress signaling pathway in cell and mouse models: The role in toxicity evaluation. *Biomaterials*, 61 (Aug 2015), p. 307.

S.B. Jaffri, K.S. Ahmad, P. Ehrh, Fabricated ZnOnanofalcates and their photocatalytic and dose dependent in vitro bioactivity. *Open Chem.*, 16 (2018), pp. 141–154. https://dx.doi.org/10.1515/chem-2018-0022.

S. Khan, M. Wajid Ullah, R. Siddique, G. Nabi, S. Manan, M. Yousaf, H. Hou, Role of recombinant DNA technology to improve life. *Int. J. Genomics*, 2016 (2016), p. 2405954.

A. Konar, S. Datta, Strain improvement of microbes, in *Industrial Microbiology and Biotechnology*, edited by Verma P., 169–193, Springer, Singapore, 2022.

M. Lajcak, V. Heidecke, K.H. Haude, N.G. Rainov, Infection rates of external ventricular drains are reduced by the use of silver-impregnated catheters. *Acta Neurochir. (Wien)*, 155, no 5 (May 2013), p. 875.

Q. Li, S. Yu, L. Li, G. Liu, Z. Gu, M. Liu, Z. Liu, Y. Ye, Q. Xia, L. Ren, Microbial communities shaped by treatment processes in a drinking water treatment plant and their contribution and threat to drinking water safety. *Front. Microbiol.*, 8 (Dec 2017), p. 2465.

S. Liao, Y. Zhang, X. Pan, F. Zhu, C. Jiang, Q. Liu, Z. Cheng, G. Dai, G. Wu, L. Wang, L. Chen, Antibacterial activity and mechanism of silver nanoparticles against multidrug-resistant Pseudomonas aeruginosa. *Int. J. Nanomed.*, 14 (2019), pp. 1469–1487. https://doi.org/10.2147/ijn.s191340.

E. Liwarska-Bizukojc, Evaluation of ecotoxicity of wastewater from the full-scale treatment plants. *Water*, 14, no. 20 (Oct 2022), p. 3345.

H. Lu, J. Wang, M. Stoller, T. Wang, Y. Bao, H. Hao, An overview of nanomaterials for water and wastewater treatment, in *Advances in Materials Science and Engineering* (2016). https://doi.org/10.1155/2016/4964828.

A. Lustosa, A. de Jesus Oliveira, P. Quelemes, A. Placido, F. da Silva, I. Oliveira, M. de Almeida, A. Amorim, C. Delerue-Matos, R. de Oliveira, D. da Silva, P. Eaton, J. de Almeida Leite. In situ synthesis of silver nanoparticles in a hydrogel of carboxymethyl cellulose with phthalated-cashew gum as a promising antibacterial and healing agent. *Int. J. Mol. Sci.*, 18 (2017), p. 2399. https://doi.org/10.3390/ ijms18112399.

C. Malarkodi, S. Rajeshkumar, K. Paulkumar, M. Vanaja, G. Gnanajobitha, G. Annadurai, Biosynthesis and antimicrobial activity of semiconductor nanoparticles against oral pathogens. *Bioinorg. Chem. Appl.*, 2014 (March 2014), p. 1.

S. Malik, S. Kishore, J. Bora, V. Chaudhary, A. Kumari, P. Kumari, L. Kumar, A. Bhardwaj, A comprehensive review on microalgae-based biorefinery as two-way source of wastewater treatment and bioresource recovery. *CLEAN - Soil Air Water* (2022). https://doi.org/10.1002/clen.202200044.

N.D. Menon, M.S. Kumar, T.G. Satheesh Babu, S. Bose, G. Vijayakumar, M. Baswe, M. Chatterjee, G.B. Kumar, A novel N4-like bacteriophage isolated from a wastewater source in South India with activity against several multidrug-resistant clinical *Pseudomonas Aeruginosa* Isolates. *mSphere*, 16, no. 1 (Jan 2021), p. e01215.

R.K. Oruganti, K. Katam, P.L. Show, V. Gadhamshetty, V.K.K. Upadhyayula, D. Bhattacharyya, A comprehensive review on the use of algal-bacterial systems for wastewater treatment with emphasis on nutrient and micropollutant removal. *Bioengineered*, 13, no. 4, (2022), pp. 10412–10453. https://doi.org/10.1080/21655979.2022.2056823.

V.D. Rajeswari, E.M. Eed, A. Elfasakhany, I. AnjumBadruddin, S. Kamangar, K. Brindhadevi, Green synthesis of titanium dioxide nanoparticles using Laurusnobilis (bay leaf): Antioxidant and antimicrobial activities. *Appl. Nanosci.*, 13 (2021), p. 1477.

Y. Ravikumar, S.A. Razack, J. Yun, G. Zhang, H.M. Zabed, X. Qi, Recent advances in Microalgae-based distillery wastewater treatment. *Environ. Technol. Innov.*, 24 (Nov 2021), p. 101839.

C. Recordati, M. De Maglie, S. Bianchessi, S. Argentiere, C. Cella, S. Mattiello, F. Cubadda, F. Aureli, M. D'Amato, A. Raggi, C. Lenardi, P. Milani, E. Scanziani, Tissue distribution and acute toxicity of silver after single intravenous administration in mice: Nano-specific and size dependent effects. *Part. Fibre Toxicol.*, 13 (Feb 2016), p. 12.

G. Rutherford, Some E. coli bacteria thrive in wastewater treatment plants: Study. *Troy Media Digital Solutions* (2021). https://troymedia.com/health/e-coli-bacteria-thriving-in-wastewater-treatment-plants/.

A. Saravanan, P. Senthil Kumar, R.V. Hemavathy, S. Jeevanantham, M.J. Jawahar, J.P. Neshaanthini, R. Saravanan, A review on synthesis methods and recent applications of nanomaterial in wastewater treatment: Challenges and future perspectives. *Chemosphere*, 307, no. 1 (2022a), p. 135713. https://doi.org/10.1016/j.chemosphere.2022.135713.

A. Saravanan, P. Senthilkumar, B. Ramesh, S. Srinivasan, Removal of toxic heavy metals using genetically engineered microbes: Molecular tools, risk assessment and management strategies. *Chemosphere*, 298 (July 2022b), p. 134341.

Y. Sun, X. Li, H. Xie, G. Liu, Removal of pollutants and accumulation of high-value cell inclusions in heavy oil refinery wastewater treatment system using *Rhodopseudomonas* and *Pseudomonas*: Effects of light intensity. *Chem. Eng. J.*, 430 (Feb 2022), p. 132586.

K.H.D. Tang, N.M. Darwish, A.M. Alkahtani, M.R. Abdel Gawwad, P. Karácsony, Biological removal of dyes from wastewater: A review of its efficiency and advances, trop. *Aqua. Soil Pollut.*, 2, no. 1 (2022), pp. 59–75. https://doi.org/10.53623/tasp.v2i1.72.

A. Tay, H. McCausland, A. Komeili, D. Di Carlo, Nano and microtechnologies for the study of magnetotactic Bacteria. *Adv. Funct. Mater.*, 29, no. 38 (2019), p. 1904178.

N.H. Torres, B.S. Souza, L.F. RomanholoFerreira, A. Silva Lima, G.N. Santos, E.B. Cavalcanti, Real textile effluents treatment using coagulation/flocculation followed by electrochemical oxidation process and ecotoxicological assessment. *Chemosphere*, 236 (Dec 2019), p. 124309.

A. Tsuda, T.C. Donaghey, N.V. Konduru, G. Pyrgiotakis, L.S. Van Winkle, Z. Zhang, P. Edwards, J.-M. Bustamante, J.D. Brain, P. Demokritou, Age-dependent translocation of gold nanoparticles across the air-blood barrier. *ACS Nano*, 13, no. 9 (Sep 2019), p. 10095.

G. Vargas, J. Cypriano, T. Correa, P. Leão, D.A. Bazylinski, F. Abreu, Applications of magnetotactic bacteria, magnetosomes and magnetosome crystals in biotechnology and nanotechnology: Mini-review. *Molecules*, 23, no. 10 (Sep 2018), p. 2438.

L.L. Wang, C. Hu, L.Q. Shao, The antimicrobial activity of nanoparticles: Present situation and prospects for the future. *Int. J. Nanomed.*, 12 (Feb 2017), p. 1227.

X. Wang, Y. Li, J. Zhao, H. Yao, S. Chu, Z. Song, Z. He, W. Zhang, Magnetotactic bacteria: Characteristics and environmental applications. *Front. Environ. Sci. Eng.*, 14, no. 4 (Aug 2020), p. 56.

G.K. Weldegebrieal, Synthesis method, antibacterial and photocatalytic activity of ZnO nanoparticles for azo dyes in wastewater treatment: A review. *Inorg. Chem. Commun.*, 120 (2020), p. 108140. https://doi.org/10.1016/j.inoche.2020.108140.

P. Wu, L. Xie, W. M.B. Wang, H. Ge, X. Sun, Y. Tian, R. Zhao, F. Zhu, Y. Zhang, Y. Wang, The biodegradation of carbaryl in soil with *Rhodopseudomonascapsulata* in wastewater treatment effluent. *J. Environ. Manage.*, 249 (Nov 2019), p. 109226.

P.R. Yaashikaa, M. Keerthana Devi, P. Senthil Kumar, Engineering microbes for enhancing the degradation of environmental pollutants: A detailed review on synthetic biology. *Environ. Res.*, 214, no. 1 (2022), p. 113868. https://doi.org/10.1016/j.envres.2022.113868.

2 Pharmaceuticals as Emerging Pollutants

An Insight on their Hazardous Impacts and Treatment Technologies

Qurban Ali, Awais Qasim, Warda Sarwar, Safia Ahmed, and Malik Badshah

2.1 INTRODUCTION

Water is essential for life and the functioning of the world is also dependent on the accessibility of fresh and clean water. It is necessary to monitor the quality of surface and groundwater because these are the primary sources of water for commercial and domestic uses. Water bodies have recently been affected by emerging contaminants that enter the ecosystem and cause harmful effects on human health and ecology. Pharmaceutical contaminants (PCs) are the byproducts or leftover drugs used to treat human or animal ailments. Antibiotics, analgesics, antifungals, anti-inflammatories, and psychiatric drugs are among the most alarming classes of emerging contaminants that rise from the pharmaceutical sectors (Samal et al., 2022; Bhushan et al., 2020). Personal care products such as detergents, cosmetics, bath soaps, lotions, dental care products, fragrances, etc. are primarily used to improve the quality of daily life (Ebele et al., 2017). PCs and personal care products are recognized as the primary sources of pharmaceutical pollutants in the environment. In the past few decades, there has been a massive increase in the production and use of pharmaceuticals and personal care products but waste management and treatment strategies could not gain much attention thus resulting in disturbing the balance. When the effluent of pharmaceutical industries is released into the water bodies, it can cause mutagenic, genotoxic, and ecotoxicological impacts on humans, animals, and plants. The presence of pharmaceuticals like estrogen in potable water can reduce male fertility and can also increase the incidence of testicular and breast cancer. Moreover, the continuous discharge of these contaminants erodes the quality of the receiving water course. Besides, active pharmaceutical composites are recalcitrant and reside for a long duration in water reservoirs and may have chronic impacts on aquatic life. The PCs are highly mobile in an aqueous environment and they get internalized by microorganisms. The recalcitration and interaction with organisms make them

DOI: 10.1201/9781003368120-2

a menace to the ecosystem. Thus, the proper treatment of polluted water is a rising challenge and the presence of PCs in water has grabbed the attention of scientists due to their low toxic dose (Samal et al., 2022). Therefore, it is important to develop efficient and effective treatment techniques for the removal of PCs from wastewater. Various techniques are being investigated for the removal of pharmaceutical pollutants from wastewater. Some of the selected treatment strategies with strong potential for large-scale use in the treatment of wastewater containing pharmaceuticals and personal care products are described below.

2.2 PHARMACEUTICAL CONTAMINANTS

Pharmaceuticals, a major milestone in human scientific advancement, have played a pivotal role in increasing life expectancy, treating millions of severe diseases, and enhancing the overall quality of life. Despite being successful, they have now emerged as dramatically increasing environmental pollutants. PCs enter the environment as a result of anthropogenic activities in the form of industrial effluents, hospital discharges, agricultural run-offs, and human and animal excreta. PCs include an array of therapeutic agents, antimicrobial cleaning solutions, and personal care products that are majorly used for treatment and disinfecting purposes. These agents are categorized into different classes based on mechanism of action, chemical structures, and treatment of diseases. Some of the major categories of pharmaceutical pollutants include antibiotics, antifungals, analgesics, anti-inflammatories, psychiatric drugs, and disinfectant solutions (Table 2.1; Samal et al., 2022). These agents are regularly detected in water samples from sewage lines, streams, ponds, and rivers as they are stable in aqueous environments and show resistance upon degradation. Moreover, PCs are target-specific compounds that have varied structures and are designed to absorb and distribute inside the human body. Their persistent presence in water bodies favors their uptake by aquatic organisms. Direct intake via gills is reported for some fish species, however, an internalization pathway varies in different species. Besides, the uptake of PCs by vegetation from the soil via roots has also been investigated (Bhushan et al., 2020). Some of the major PCs are discussed below.

TABLE 2.1
Class of Pharmaceuticals, Uses, and Adverse Impacts on Humans

Class of Pharmaceuticals	Uses	Adverse Impacts on Human
Antibiotics	Treat bacterial infection	Liver dysfunction, Cardiac diseases, Immune disruption, Bone-marrow suppression
Anti-cancer drugs	Treat cancers	Embryo toxic and teratogenic for the fetus
Antidepressants	Treat mental illnesses	Hypoglycemia, sexual dysfunction, growth retardation
Analgesics & Anti-inflammatory	Painkillers & antipyretic	Cardiac diseases, hypertension, mental health issues

2.2.1 ANTIBIOTICS

The need for antibiotics is inevitable for a healthy life. Since the discovery of the first antibiotic, penicillin, the demand for novel and effective antibiotics is ever growing. B-lactams, aminoglycosides, macrolides, fluoroquinolones, sulfonamides, and lincomycin are the major classes of antibiotics in clinical practice (Kumar et al., 2019). Along with clinical use, a sub-therapeutic dose of antibiotics is added to the feed which acts as a growth promoter for poultry chickens and livestock. Studies revealed that mild doses of antibiotics prevent covert infections; and reduce bacterial-associated molecular patterns thereby averting immune responses. It is also believed that changes in the intestinal microbiota of farm animals and birds result in lower competition among bacterial species for nutrients and food absorption is enhanced which in turn increases muscle mass (Castanon, 2007). The mounting demand for pharmaceutical products worldwide boosted their production and increased their likelihood to accumulate in the ecosystem. All the above-mentioned antibiotics bind to a specific target but over time, these products are exposed to environmental heterogeneity. Consequently, structural changes occur and their binding specificity is compromised. With the altered specificity in the ecosystem, it is more likely to affect a wide range of organisms by binding with their cellular components. In humans, nearly 90% of the soluble and insoluble antibacterial agents leave the body through urination and defecation, respectively (Hassan et al., 2019). Hence, in this way, these antibacterial agents get into the receiving water course. Several studies have reported a significant amount (nano to micrograms per liter) of these agents in surface and groundwater sources. Bacterial flora is exposed to these agents which makes wild-type bacteria resistant by acquiring resistance mechanisms (Larsson, 2014). The exchange of antibiotic resistance genes (ARGs) among species via horizontal gene transfer favors the ARGs' dissemination. The resistance poses a threat to healthcare by enhancing infection rates in humans and animals and has a devastating impact on the economy. Antibiotic resistance-associated mortalities and morbidities are one of the top leading causes of death globally (Zhang et al., 2020). Furthermore, PCs have already contaminated the groundwater resources to a great extent and incorporated into our daily diet (Wilkinson et al., 2022).

2.2.2 ANTIFUNGAL AGENTS

Antifungal agents are a class of antimicrobial drugs used for the treatment of fungal infections. Examples include polyenes, echinocandin, azoles, and nucleoside analogs. Polyenes and azoles act on the fungal cell membrane by binding to the ergosterols while echinocandin is a cell wall synthesis inhibitor. The nucleoside analogs inhibit the synthesis of nucleic acids by replacing nucleosides in the growing chain of nucleic acid. Their selective nature allows them to be target-specific with less toxicity for the host. However, the excessive use of antifungal agents has caused the emergence of resistance in fungi. The resistant strains are responsible for an amplified rate of infections and hospitalization.

Azole is a major class of antifungals that consists of two subclasses: imidazoles and triazoles. Examples of imidazoles are clotrimazole, econazole, miconazole, ketoconazole and metronidazole while sub-types of triazoles are isavuconazole, voriconazole, fluconazole, posaconazole and itraconazole. The presence of azoles has been reported in several studies. Advanced detection techniques such as liquid chromatography-mass spectrometry (LC-MS) have made it possible to detect a minute quantity of drugs in water samples from the environment (Abdallah et al., 2019; Monapathi et al., 2021). Specifically, fluconazole is usually detected in water samples as compared to other antifungal agents due to its wide use and hydrophilic nature. In contrast, the hydrophobicity of other antifungals encourages adsorption onto solids.

2.2.3   ANALGESICS AND ANTI-INFLAMMATORY DRUGS

Analgesics and anti-inflammatory drugs are used to relieve pain, inflammation-associated discomfort, headache, and pyrexia. A list of analgesics and anti-inflammatory drugs includes; diclofenac, ibuprofen, ketoprofen, cannabis, opioids, alcohol, paracetamol, etc. Among the mentioned drugs, paracetamol, alcohol, ibuprofen, and diclofenac are the widely used drugs. Above 30 million dosages of non-steroidal anti-inflammatory drugs are consumed on daily basis around the globe (Feng et al., 2013). The same routes as antibiotics are employed for the release of analgesics from the body (Ahmed et al., 2021a). Even for low concentrations of anti-inflammatories, toxicity toward living systems is documented. The ever-growing demand for analgesics is creating an atmosphere that gives way to its accumulation in the environment. Continuous production and discharge require a sophisticated system for the treatment of pharmaceutical and cosmetics wastes.

2.2.4   PSYCHIATRIC DRUGS

Psychiatric medicines contain neuro-active complexes which are used to treat depression and other psychiatric conditions (Escudero et al., 2021). These drugs normalize the imbalance of neurotransmitters with good absorption and the bio-availability is 60%–100%. Several studies have reported the detection of psychiatric agents in water bodies (surface, sewage, and wastewater). The lethal impact of these pharmaceutical products is well documented. Antidepressants cover the major portion of psychiatric medicines (Mamta et al., 2020). Some examples are selective serotonin reuptake inhibitors (SSRIs), serotonin modulators and stimulators (SMSs), serotonin-norepinephrine reuptake inhibitors (SNRIs), nor-epinephrin reuptake inhibitors (NRIs), etc. (Rosenblat and McIntyre, 2020).

2.2.5   DISINFECTANTS

Healthcare facilities are home to diverse microorganisms, especially pathogenic species. In order to prevent the spread of these pathogenic microbes, certain chemicals called disinfectants are applied to contaminated sites to reduce the microbial load. In the recent past, during the COVID pandemic, the frequent use of chlorine-based sprays in hospitals, houses, community centers, mosques, and markets has boosted the level of environmental disinfectant residues. Some reports depicted 0.4 mg/L chlorine residues in water samples from Wuhan Lake at the start of the pandemic, showing the direct relation of the pandemic with disinfectant pollution (Chu et al., 2020). The chemical structure of these disinfectants is compromised in water and free chlorine is produced. Upon reaction with naturally present organic materials in water bodies, the free chlorine results in the formation of byproducts including trihalomethanes, trihalophenols, haltoacetonitrils, haloacetic acids, and haloketones (Parveen et al., 2022).

2.3 SOURCES OF PCs

Residual pharmaceutical products can enter the aqueous environment through a variety of direct and indirect routes. Figure 2.1 shows the key sources of these emerging contaminants of concern. PCs are released into water streams primarily via inappropriate disposal, and discharge of untreated or inadequately treated pharmaceutical wastewater into main water channels. Moreover, the waste and effluent from hospitals and healthcare centers also release pharmaceuticals into the environment. Besides, the majority of antibiotics are not completely metabolized by humans and animals and a high percentage of administrated antibiotics are released into the environment via urination and defecation. Similarly, low doses of antibiotics are also used as a medicine as well as additives to animal feed in livestock and aquacultures. Such continuous use of pharmaceutical agents results in the persistent release of leftover complexes in the form of fecal, urinary, and washing area wastewater into the water bodies. The agriculture sector is also responsible for the release of these

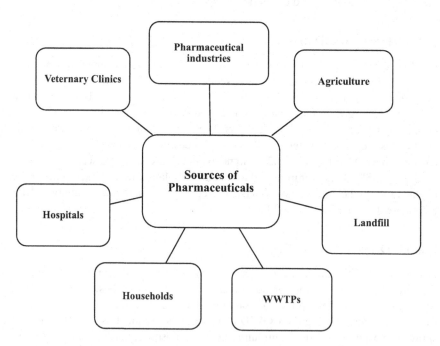

FIGURE 2.1   The figure showing the various sources of pharmaceutical contaminants enter into the environment.

types of contaminants into the environment as various agrochemicals (pesticides and insecticides) are usually sprayed on the crops thus resulting in their introduction into the soil where these pharmaceutical pollutants get percolated or leached down or run off through surface into the water reservoirs. Moreover, treated cadavers and disposed of fallow drugs are also a source of contaminants for the aquatic ecosystem (Eniola et al., 2022).

2.4 IMPACT OF PCs ON AQUATIC LIFE

PCs enter into water bodies through various routes discussed above. Although the existence of pharmaceuticals in the aquatic system may be in low concentrations, their widespread use, low degradation, high reactivity with the biological environment, and constant discharge in water bodies make them pseudo-persistent in aquatic systems. The continuous discharge of pharmaceutical pollutants into water bodies as well as their exposure may have chronic impacts on aquatic life. Such as the presence of tetracycline in the aquatic system makes fishes immune-compromised. It is also reported that the existence of macrolides enhances the deformity of the yolk sac edema, and uninflated swim bladder, and also affects the movement of the embryo in zebrafish. Moreover, it is also reported that exposure to rapamycin even at very low concentrations can affect the behavior of zebrafish. Likewise, sulfamethoxazole and norfloxacin have hazardous impacts on the growth and reproduction of zebrafish. Pharmaceuticals such as tetracycline and fluoroquinolones can also affect the behavior and development by changing the expression of various genes in zebrafish (Chen et al., 2014; Kotwani et al., 2021).

High levels of estrogen cause vitellogenesis in male *Oryzias latipes* and also enhance the mortality rate in fishes. The occurrence of estrogen in water streams changes male fish into female fish through the development of female traits (Samal et al., 2022). The existence of pharmaceutical pollutants in the aquatic microcosms, wastewater, natural marine, and freshwater affects the species of aquatic animals as well as microbial diversity and their functions at various tropic levels. Both chronic and acute adverse impacts have been observed on aquatic algal communities in the existence of different antibiotics such as tetracycline, quinolones, macrolides, and sulfonamides. Photosynthesis in cyanobacteria is also affected by tetracycline and macrolide with these contaminants also disrupting the balance between beneficial bacteria and poisonous weeds thus enhancing the population of the latter. Crustaceans also show chronic toxicity because of quinolones and macrolide accumulation (Shakoor et al., 2020).

According to an environmental risk assessment study involving around 226 antibiotics, 44% and 20% of antibiotics are extremely poisonous to daphnids and algae, respectively (Bilal et al., 2020). A variety of antibiotics, including tetracycline, macrolide, and sulfonamides exhibit harmful impacts on the development and growth of algae. Such as, long-time exposure to sulfathiazole may cause growth retardation in macroalgae *Lemnagibba*, while tetracycline, aureomycin, and oxytetracycline have shown significant impacts on the growth of *Microcystis aeruginous* (Shang et al., 2015). Furthermore, the presence of tetracycline in the aquatic environment causes histological changes in *Gambusia holbrooki*, such as hepatocellular vacuolization and sinusoidal enlargement (Nunes et al., 2015). Even a low concentration of oxazepam, an anti-anxiety drug, in the aquatic environment, can considerably affect the fish's behavior and rate of food intake (Kayode-Afolayan et al., 2022).

2.5 IMPACT OF PCs ON HUMAN HEALTH

The continuous discharge of untreated pharmaceutical wastewater into main water bodies like rivers has an adverse impact on the environment as this water may be used for irrigation purposes, restoration, industries, groundwater sources, and potable and non-potable municipal uses. This water may contain various antibiotic-resistant bacteria, chemicals, and pharmaceutical residues that pose risks to humans, animals, and the environment if not mitigated properly. Various classes of antibiotics including sulfonamide, quinolones, macrolide, ciprofloxacin, and chloramphenicol may persist in portable water for months, and many are not eliminated completely by traditional disinfection methods in drinking water treatment systems (Kotwani et al., 2021). On the other hand, PCs can also intercalate into the food chain, via plants, when the water containing pharmaceutical pollutants is used for irrigation. As a result, the presence of the above-mentioned pharmaceuticals in drinking water or food can affect the health of humans and animals. Major health concerns linked with exposure to antibiotics are liver dysfunction, heart problems, compromised immune system, bone marrow suppression, and intercalation in the food chain. Likewise, antidepressants cause hypoglycemia, sexual dysfunction, and retardation of growth in aquatic organisms as well as in humans (Bhushan et al., 2020). Furthermore, residues of anti-cancer therapeutics in potable water are specifically hazardous during pregnancy because they have been shown to cross the placental membrane where they induce embryotoxicity and teratogenicity in the fetus (Samal et al., 2022). The hydrophobicity of several PCs and cosmeceuticals enhances their potential to accumulate in high-fat-containing tissues thus stimulating endocrine disruption in humans and animals (Ahmed et al., 2021a). Along with physiological consequences in humans and animals, the presence of antibiotics in the environment promotes the emergence of antibiotic-resistant bacteria (ARB), carrying the antibiotic-resistant gene (ARGs). The acquisition of ARGs by pathogenic bacteria causes a great risk to human and animal health (Shakoor et al., 2020).

2.6 IMPACT OF PCs ON AGROECOSYSTEM

The effect of PCs on the agroecosystem can be best understood from the ecological risk assessment of the influence of pesticides on alluvial soil, plant communities, and associated aquatic habitats. The presence of antibiotic residues—the plant toxins—in soil alters the microbial populations thus affecting the plant rhizosphere structure that consequently leads to low agroecosystem production with considerable agronomic impacts. Moreover, the uptake of pharmaceuticals by plants has been reported to affect plant growth and development, root alterations, changes in reproduction rate, and reduction in phosphorus assimilation. However, different factors affecting the toxicity of pharmaceuticals towards plants include the type of pharmaceutical pollutant and plant species, and environmental conditions (Shakoor et al., 2020; Łukaszewicz et al., 2016).

Some of the plausible mechanisms stated to be involved in the uptake of pharmaceutical pollutants by plants include ion tapping and protein-mediated transport but presently the evidence supporting these assertions is scarce. However, it is evident that transpiration—a passive mechanism—is the primary process for the uptake of pharmaceuticals by plants. As PCs are a diverse group of chemical substances, their interaction with plants is presumed to be structure-dependent. It has been investigated that ionizable chemicals have restricted uptake as they are mostly ionized in the natural environment and are sorbed by the soil minerals. Similarly, hydrophobic pharmaceuticals also have limited mobility and hence congregate in the roots. Moreover, it has also been reported that acidic or neutral compounds have a greater transport rate through transpiration compared to basic chemicals. Apart from this, the uptake of small molecules has been found to be linked to translocation through water mass flow thus localizing the pharmaceuticals in older leaves. But this is not a universal rule for all vegetation (Łukaszewicz et al., 2016).

2.7 IMPACT OF PCs ON MICROBIAL LIFE

The unprecedented use of antibiotics for therapeutic and prophylactic purposes and the presence of PCs in the environment have resulted in increased resistome (resistance gene pool), direct tissue toxicity, and modulations in microbiomes. In the aquatic ecosystem, the emergence of antibiotic-resistant bacteria and antibiotic-resistant genes has emerged as a major environmental health concern (Shakoor et al., 2020). Humans, animals, and the environment serve as reservoirs for AMR, allowing it to disseminate through linked ecosystems. One of the biggest environmental drivers in propagating AMR includes the pharmaceutical industries, which are expanding extensively to supply the ever-increasing demand for antibiotics. The pharmaceutical effluent contains a cocktail of antibiotics and antibiotic resistance genes (ARGs), and these facilities serve as hotspots for environmental pollution and the propagation of AMR. Antibiotic resistance genes can either spread through horizontal or vertical gene transfer. Both of these phenomena co-exist naturally in the environment. Moreover, co-resistance and cross-resistance also stimulate the co-selection of resistance genes, plasmids, mobile genetic elements, and virulence factors (Kotwani et al., 2021).

Insufficient treatment of pharmaceutical effluents and their unregulated disposal are the major sources of antibiotic pollution and their persistence in the environment. Antimicrobial residues present in the effluent affects the environmental microbiome at varying concentrations. Sub-inhibitory concentrations of pharmaceutical pollutants have been reported to cause stunted growth in susceptible bacteria where antibiotics function as signaling molecules thus modulating bacterial genome expression, gene transfer, quorum sensing, biofilm formation, and virulence. It has also been investigated that long exposure to sub-inhibitory doses of antibiotic residues causes the induction of bacterial SOS repair system, frequency of genomic mutations, formation of mobile genetic elements, transfer of genes, and over-expression of some genes responsible for the propagation of AMR (Martínez, 2017). Besides water bodies, the pharmaceutical effluents are also expelled into the environment via soil thus instituting soil as a vector for antimicrobial resistance genes and a reservoir for antibiotic residues. Studies have revealed the presence of PCs in the soil that results in structural alterations in the soil microbiome resulting in decreased biomass and microbial activities including respiration, nitrification, and denitrification. Furthermore, these pollutants may also affect the bacterial enzymatic activity such as ureases, dehydrogenases, and phosphatases that are reckoned as the major indicators of soil activity thus causing the loss of important microbial communities crucial for ecological events (Cycon et al., 2019).

2.8 DETECTION METHODS OF PCs IN WASTEWATER

The rise in reported detection of PCs at minute concentrations in different environmental matrixes, such as water cycle (e.g., groundwater, surface water, potable water, treated wastewater effluent) is primarily referable to technological advances in the accuracy and sensitivity of detection instruments and analytical techniques. Liquid chromatography with mass spectrometry (LC-MS) or tandem mass spectrometry (LC-MS/MS)[2] and Gas chromatography with mass spectrometry (GC-MS) or tandem mass spectrometry (GC-MS/MS)[2] are advanced tools that can determine target molecule at the concentration of ng/L level and are widely used for the detection of pharmaceuticals in wastewater and potable water. The selection of techniques is based on the chemical and physical features of targeted pharmaceuticals. For pharmaceuticals that are polar and soluble in water LC-MS/MS analytical technique is suitable. However, for pharmaceuticals that are volatile in nature GC-MS/MS is a more suitable analytical technique. Moreover, high-performance liquid chromatography is another technique that is used to detect a variety of PCs from different samples (Samal et al., 2022; WHO, 2012).

2.9 TREATMENT STRATEGIES

The pharmaceutical sector uses different types of wastewater treatment systems. Wastewater produced by pharmaceutical industries differs not only in quantity but also in composition, plants, season, and even period of time, depending on methods and raw materials utilized in the production of different medications. The location of the production plant also brings in a variable related to the quality of accessible water. Therefore, it is challenging to specify a specific treatment method for such a diversified sector. Lots of alternative treatment techniques are accessible to deal with a wide range of waste generated by pharmaceutical industries (Gadipelly et al., 2014). Generally, treatment methods in this sector fall into three types: physical, chemical, and biological methods (Figure 2.2).

2.9.1 PHYSICAL TREATMENT STRATEGIES

The elimination of contaminates from wastewater without modifying their biochemical properties is known as the physical wastewater treatment method. Such treatment techniques neglect the effect of any biological or chemical agents. Physical methods are frequently used before chemical and biological wastewater treatment processes. Physical wastewater treatment techniques include pretreatment methods (screening,

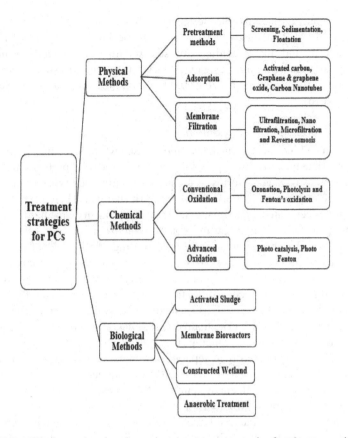

FIGURE 2.2 The figure showing the various treatment strategies for pharmaceutical industries wastewater.

sedimentation, and floatation), adsorption, and membrane filtration. These physical treatment methods are based on a mass transfer strategy (Samsami et al., 2020). The major benefit of adopting physical treatment methods is that they are technologically basic and versatile because they use simple equipment and can be adapted to a variety of treatment processes (Crini and Lichtfouse, 2019). Furthermore, when compared to other methods, the production of solid waste is considerably lower (Ahmed et al., 2021b).

2.9.1.1 Physical Pretreatment Methods

The most considerable pretreatment methods in wastewater treatment techniques include screening, sedimentation, and floatation (Ahmed et al., 2021b).

2.9.1.1.1 Screening

Screening refers to the removal of larger and floating solid objects that may clog subsequent treatment strategies, it is also known as one of the oldest techniques of physical treatment of wastewater (Naidoo and Olaniran, 2014). The screening is necessary for wastewater treatment to ensure the proper operation of various processes by preventing possible destruction, blockage, and process disruption. Screening is divided into two types, based on the kind of impurities removed: coarse screening and fine screening. Coarse screening mechanically or manually eliminates particles (rags, plastics, debris) larger than 6 mm in size, however, suspended solid particles having a size of 0.001–6 mm are removed by fine screening (Bhargava, 2016).

2.9.1.1.2 Comminution and Sedimentation

Comminution and sedimentation are commonly used where screening is problematic or not appropriate for a given setup. In these methods, the solid particles are smashed into fragments using several forms of crushers. Usually, such types of crushers are placed between the sedimentation tank and the grit chamber. Thereafter, the crushed particles are precipitated and separated from the effluent (Chauhan and Kumar, 2020). The grit chamber and settling container assist in the isolation of floating particles. A grit chamber aims to ensure the settling of the residual large materials so that these particles do not block the wastewater treatment plant. Based on their qualities, subject of interest, and size, the grit chambers have different types which include aeration chamber, vortex flow chamber, and horizontal flow chamber (Esfahani et al., 2018). These various kinds of grit chambers start sedimentation in various manners. Besides the grit chambers, sedimentation containers are placed to ensure that the wastewater is cleaned properly. The sedimentation container allows the slurry to settle down and then separates the precipitated denser particles from the liquid phase. The placement of these settling containers is usually coupled with a biological wastewater treatment system, in which dense sediment mixed with biological bodies is separated from wastewater in a sedimentation container (Naidoo and Olaniran, 2014). Gravitational force is used in sedimentation containers to settle down heavier materials. Depending on their configuration, settling tanks can be classified into three types: inclined surface (depth separated into shallower parts for quick settlement), solid contact (the liquid rises upward and hard particles remain in the settling tank), and horizontal flow (velocity and flow distribution are constant; Pickering and Hiscott, 2015). The settling time and velocity of wastewater are the elements that affect the rate of sedimentation in each of these distinct sedimentation techniques (Ramin et al., 2014).

The sedimentation in the settling container or grit chamber aids in the startup of subsequent treatment. Furthermore, the idea of an equalization basin is crucial in this respect as it tends to boost the effectiveness of the secondary and tertiary stages of the wastewater treatment system by decreasing the flow rate and contaminant level and by regulating the temperature of the system (Verlicchi et al., 2015).

2.9.1.1.3 Flotation

Except for sedimentation, flotation can also remove suspended solids of secondary effluent. The technology characteristic involves producing a large number of tiny bubbles by injecting air into wastewater, forming a floating floc with a smaller density than wastewater. It then floats to the surface of the wastewater where it is separated (Thapa et al., 2022).

2.9.1.2 Adsorption

Adsorption is the most prominent physical wastewater treatment method, which is frequently employed for the elimination of trace organic contaminants from water and as the key mechanism for eradicating pharmaceutical waste from pharmaceutical wastewater (Wang and Wang, 2016). A number of inorganic, organic, and hazardous pollutants can also be removed using an adsorption technique. Temperature, size, concentration molecular mass, and other chemical properties of pollutants can affect the adsorption efficiency of adsorbents (Lima, 2018). To improve the pharmaceutical waste adsorption capability, various adsorbents have been designed and investigated for the adsorption of waste from pharmaceutical industries' wastewater. This part will focus on carbon-based adsorbent tools such as activated carbon, carbon nanotubes, and graphene.

2.9.1.2.1 Activated Carbon

Activated carbon, a conventional adsorbent, is often employed in wastewater treatment and has a specific application for pharmaceutical waste from a liquid suspension. There are two different forms of activated carbon which include granular activated carbon and powdered activated carbon. Powdered activated carbon offers advantages over granular activated carbon since it is normally fresh, whereas granular activated carbon is generally recycled in fixed bed columns. Powdered activated carbon was able to partly eliminate all the target pollutants in diverse wastewater sources, based on the physicochemical features of each contaminant. Because of the high surface area of activated carbon (about $1,000\,m^2/g$) and the combination of well-designed pore structure and surface chemical features, activated carbon adsorption is well-suited to eliminate pharmaceutical waste (Gadipelly et al., 2014). However, two complications occur with an increase in operational time. One is the reduction in the adsorption efficiency of activated carbon, and the other is the degradation of activated carbon in a complicated wastewater treatment plant. Further research is required to overcome these challenges. The absorption efficiency of activated carbon to pharmaceutical waste primarily depends upon the charge and hydrophobicity of the contaminant (Wang and Wang, 2016). Along with the charge and hydrophobicity of PCs, laboratory-scale, and pilot-scale investigations have revealed that the water matrix also influences the pharmaceutical pollutant's adsorption by activated carbon. In a pilot-scale investigation using powdered activated carbon as an absorbent, it was revealed that organic contaminant present in the wastewater can compete with pharmaceutical waste for binding sites of activated carbon, which reduced the adsorption ability of activated carbon to PCs (Mailler et al., 2015). Furthermore, the elimination performance of pharmaceutical pollutants by activated carbon is influenced by a variety of parameters, including pH, the structure of activated carbon, and contact time (Nam et al., 2014; Wang and Wang, 2016).

2.9.1.2.2 Graphene and Graphene Oxide

Graphene is a unique polymer with a single chip form made up of carbon atoms. The basic structure of graphene is a two-dimensional array of carbon atoms chemically attached through sp^2 hybrid orbitals to create a honeycomb sheet. Graphene oxide, a precursor of graphene, is always formed by the oxidation of graphite. Graphene and its oxide have recently gained increased consideration in pharmaceutical wastewater treatment because of their extraordinary characteristics. However, the removal potential of graphene and its oxide alters with the physicochemical features of PCs. Just like with activated carbon, contact time and pH have a noticeable impact on the efficacy of graphene and its oxide in eliminating pharmaceutical waste from wastewater (Yang and Tang, 2016). Because graphene and graphene oxide have greater surface area than activated carbon, it is fair to think that graphene and its oxide could be highly effective adsorbents for the elimination of pharmaceutical pollutants. The majority of investigations on the adsorption of pharmaceutical waste by graphene and graphene oxide were conducted as batch experiments on the laboratory scale. Secondly, the solution was generally limited to simulated wastewater, which is less complex than real pharmaceutical wastewater. So, further research is needed to evaluate the efficiency of graphene and its oxide for PCs' adsorption in real pharmaceutical wastewater, the water matrix on the absorption of pharmaceutical waste, and the relationship between graphene and graphene oxide and pharmaceutical pollutants (Wang and Wang, 2016).

2.9.1.2.3 Carbon Nanotubes

Carbon nanotubes, like graphene, have exceptional features that make them a promising choice for a variety of applications such as medical devices and energy storage. Several investigations have been conducted to examine the elimination of PCs using carbon nanotubes such as carbamazepine, ketoprofen (Liu et al., 2014), triclosan (Cho et al., 2011), sulfamethoxazole, ibuprofen, caffeine, prometryn, and acetaminophen (Wang and Chu, 2016). According to these findings, carbon nanotubes exhibit a remarkable adsorption capacity to PCs. However, the adsorption potential varies depending on the surface chemistry and characteristics of carbon nanotubes. Additionally, the physicochemical features of pharmaceutical waste can affect the adsorption properties of carbon nanotubes. The carbon nanotubes that have bigger inner diameters cannot display better efficiency of adsorption with the PCs in wastewater (Wang and Wang, 2016).

The adsorption of pharmaceutical pollutants onto activated carbon, graphene and its oxide, and carbon nanotubes is believed to be a useful technique. However, the various challenges should be resolved in the next step before the above-mentioned materials may be used on large scale.

a. The ability of activated carbon to absorb macromolecules should be enhanced. Because of the steric action of the micropore, activated carbon has limited efficiency for the adsorption of macromolecules. Furthermore, research is required to reduce the steric impact of micropores.

b. It is important to reduce the production cost of graphene as its high cost restricts its applicability. As a result, more investigation should be performed in order to produce high surface area graphene at a reasonable cost.

c. It is crucial to improve carbon nanotube production technologies thus requiring further research to establish a simple and effective way of producing carbon nanotubes.

Similarly, the recycling and regeneration of activated carbon, graphene, graphene oxide, and carbon nanotubes should also be taken into consideration. Additionally, the coexistence of adsorbents and pharmaceuticals and personal care products (PPCPs) may have a hazardous impact on aquatic microorganisms, and additional research should be done to explore the association of adsorbents and pharmaceutical waste, as well as their environmental fate, including cytotoxicity and hazard in aquatic microorganisms.

2.9.1.3 Membrane Filtration

Membrane filtration is a physical wastewater treatment method that removes pollutants depending on their size and characteristics through the membrane. The hydrostatic pressure around the membrane is the primary driving force behind the filtration via the membrane (Ricceri et al., 2021). In membrane separation, a part of wastewater termed as permeate travels across the membrane, whereas particles bigger than the pore size of the membrane are excluded. Membrane-based contaminant separation is a common technique for pollutant elimination but regular modifications, inclusions, and exclusions are being made to achieve more effective contaminant separation (Obotey Ezugbe and Rathilal, 2020). Membrane filtration can be of different kinds based on the pore size of the membrane. Such as microfiltration has a pore size between 0.001 and 0.1 μm, ultrafiltration has a pore size that is less than 0.001 μm, and the pore size for nanofiltration is 1–10 nm, and so on. Furthermore, in the scenario of reverse osmosis, the use of a semipermeable membrane facilitates the elimination of particulates smaller than 1 nm (Ahmed et al., 2021b). Pharmaceuticals, in general, have a molecular mass of more than 250 Da that can be retrieved employing appropriate membrane filtration. All the above-mentioned membrane types can provide significant economic benefits.

2.9.1.3.1 Microfiltration

Microfiltration is one of the earliest commercially available pressure-driven membrane technologies and is able to eliminate contaminants having a size of μm. Since its beginning in the nineteenth century, the microfiltration membrane method has evolved into advanced technologies for cleaning wastewater carrying a variety of contaminants (Anis et al., 2019). Microfiltration membranes can be employed alone or in combination with other treatment techniques such as advanced oxidation which can be accomplished by integrating chemical oxidants to eliminate pharmaceutical waste more efficiently (Ba et al., 2018). A study was conducted in which a fabricated microfiltration membrane can remove recalcitrant pharmaceuticals from wastewater through electro-oxidation. The efficiency of electro-oxidation was enhanced by integrating G/SnO_2 into the membrane filter. It was observed that the microfiltration membrane effectively removes sulfamethoxazole with an elimination efficiency of 85% when G/SnO_2 is incorporated into the membrane filter (Yu et al., 2020).

2.9.1.3.2 Ultrafiltration

Ultrafiltration is a membrane separation process that uses low pressure to eliminate organic contaminants from wastewater. This type of membrane filtration removes different types of endocrine-disrupting compounds/pharmaceuticals about 75%–100%

from various types of water such as potable water and pharmaceutical wastewater. Sometimes, the efficiency of ultrafiltration membrane separation is reduced because of membrane fouling, thus reducing the retention capability of the ultrafiltration membrane (Eniola et al., 2022).

2.9.1.3.3 Nanofiltration

The nanofiltration membrane technology is used to eliminate different types of micro-contaminants. This type of membrane separation method uses a pressure gradient as the driving force (Eniola et al., 2022). It is the latest developed membrane filtration technique, and its applications have been growing quickly over the past decade. It has been frequently employed in the wastewater treatment system, such as the separation of antibiotic concentration from aqueous solution (Gadipelly et al., 2014). Nanofiltration removes different types of pharmaceuticals like diclofenac, ibuprofen, dipyrone, acetaminophen, and caffeine from wastewater (Eniola et al., 2022). Recovery of amoxicillin, for instance, is crucial because of its physical properties and exposure to the environment. Amoxicillin having a molecular weight of 365.40 Da is a widely used antibiotic. As a result, the amount of amoxicillin discharged into the wastewater and drinking water could be quite considerable. Nanofiltration can be employed to recover and remove amoxicillin from pharmaceutical effluent, in order to reduce the harmful impact of amoxicillin in environment. Studies have revealed the removal of amoxicillin from pharmaceutical effluent using nanofiltration membrane technology. The rejection of amoxicillin by nanofiltration was acceptable and exceeding to 97% whereas COD obtained a maximum of 40% rejection. The consistent penetration flow and strong rejection of amoxicillin suggested that nanofiltration could be used to retrieve amoxicillin from pharmaceutical effluent (Shahtalebi et al., 2011).

2.9.1.3.4 Reverse Osmosis System

A reverse osmosis system is usually used in the production of potable water from brackish water and seawater. However, it has also been applied to treat wastewater from pharmaceuticals and other industries (Eniola et al., 2022). Reverse osmosis membrane technology in various combinations has demonstrated the effective elimination of 36 personal care products and endocrine-disrupting agents, including hormones, antibiotics, antiepileptics, analgesics, and contraceptives. Most of the items studied were eliminated by the reverse osmosis membrane method to amount below the detection limit. Another research used reverse osmosis and nanofiltration methods to remove a variety of medications such as ketoprofen, carbamazepine, diclofenac, hydrochlorothiazide, and propyphenzone from full-scale potable water treatment system, with significant percent rejection of $\geq$85% for all pharmaceuticals mentioned. Both reverse osmosis and nanofiltration treatment systems have the potential to efficiently eliminate drugs from wastewater but the removal of remaining sludge containing the concentrated form of contaminants remains a challenge (Deegan et al., 2011).

2.9.2 Chemical Treatment Strategies

Chemical treatment is the technique of cleaning wastewater by applying chemicals in a series of reactions. The soluble pollutants in wastewater are driven to dissociate in chemical wastewater treatment systems by introducing particularly targeted material. In certain scenarios, biological and mechanical wastewater treatment systems are insufficient to allow treated water to access bodies of water, therefore chemical wastewater treatment technologies are useful. Chemical treatment is an essential phase in the treatment of various industrial and agricultural wastes as the pollutants need to be treated further. The chemical oxidation method, a chemical treatment system, has arisen as an innovation in which chemical oxidants are applied to transform contaminants into nontoxic and manageable forms. Chemical oxidation treatment technology is further divided into two categories: conventional oxidation and advanced oxidation (Ahmed et al., 2021b).

2.9.2.1 Conventional Oxidation Methods

2.9.2.1.1 Ozonation

Ozonation is the technique of introducing ozone into water. This is typically accomplished by releasing ozone from the bottom of a container using a sparger. Ozone exerts oxidation impacts by direct ozone reaction or indirect radical reactions via a series of oxidative processes. For certain molecules, ozone is a potent oxidant on its own. However, in the presence of water, ozone can interact with accessible hydroxide ions to generate the less selective and so more potent oxidizing hydroxyl radical. Ozone can interact with electron-rich aromatic pharmaceuticals such as ciprofloxacin, azithromycin, sulfamethoxazole, carbamazepine, clarithromycin, diclofenac, erythromycin, metoprolol, etc. and with the pharmaceuticals having deprotonated amines including trimethoprim, at a lower pH. Besides, hydroxyl radicals because of their non-selective and reactive nature oxidize the pharmaceuticals resistant to ozone under basic conditions like penicillin, cephalexin, loperamide, and primidone (Ghazal et al., 2022).

Ozone owing to its remarkable oxidation potential has recently gained significant attention as a tertiary treatment approach compared to conventional oxidants for the treatment of wastewater contaminated with pharmaceuticals. However, one of the common limitations of introducing ozonation in wastewater treatment procedures involves the presence of a disproportionate concentration of organic carbon and other oxidizable compounds thus requiring a considerable amount of ozone for the complete treatment of typical sewage (Farzaneh et al., 2020). Moreover, excessive ozone in water needs to be removed upon completion of the treatment process because of the deleterious nature of ozone. Another serious concern associated with the use of ozonation for the treatment of pharmaceutical wastewater is the formation of toxic refractory byproducts that may have toxicity higher than the parent compounds (Tufail et al., 2021).

2.9.2.1.2 Photolysis

Photolysis is a recently adopted technique for the tertiary treatment of wastewater. It involves the use of high-energy radiation like UV that destroys the pollutants by the production of strong oxidizing agents. When subjected to irradiation (UV or visible light), the pollutants (pharmaceuticals) may be excited to the singlet-excited state, which then crosses to the triplet excited state via the photolysis degradation pathway. At this stage, the molecule either reacts to create the mineralized byproducts or returns to the ground state due to the formation of hydroxyl radicals or the presence of dissolved oxygen. However, the use of photolysis for the treatment of wastewater containing pharmaceuticals has not been exploited entirely but recent research highlight the potential of using ultraviolet radiations or visible light in conjunction with powerful oxidants including O_3 or H_2O_2 to stimulate the free radical production for the decontamination of pharmaceuticals. Moreover, photolysis using UV or solar light can be used to carry out the significant degradation of pharmaceutical compounds including carbamazepine (60%), fluoxetine (57%), trimethoprim (50%), and atenolol (57%) majorly by the oxidation of hydroxyl radicals (Eniola et al., 2022).

Photolysis is recognized to be a very advantageous process since it does not involve the use of oxidizing agents or other catalysts thus making it a cost-effective technique. Nonetheless, this traditional oxidation process has certain critical shortcomings such as the organic compounds that behave like photosensitizers may cause increased turbidity in the media, obstructing UV light penetration through contaminated water consequently making photolysis a less efficient technique (Cuerda-Correa et al., 2019).

2.9.2.1.3   Fenton's Oxidation Treatment

Fenton's oxidation treatment, a heterogeneous catalytic reaction, involves the reaction of ferric ions or ferrous with hydrogen peroxide through a free radical chain reaction resulting in the production of hydroxyl radicals. Since iron serves as a catalyst for this reaction and is an abundantly present element so this method is considered to be most feasible for wastewater treatment as it can potentially decrease the toxic burden of pharmaceutical effluent and make them more susceptible to biological post-treatment. Studies have revealed 95% removal of COD in chloramphenicol and paracetamol effluents and complete degradation of penicillin upon Fenton/UV treatment for 40 minutes. Nevertheless, some of the drawbacks associated with the Fenton process include the excessive production of iron sludge and pH dependency. So, the Fenton process is most effectively used as a pretreatment procedure for transforming non-biodegradable pharmaceutical wastewater into biodegradable effluent and hence making biological effluent treatment more productive (Gadipelly et al., 2014).

$$Fe^{2+} + H_2O_2\ Fe^{3+} + OH\bullet + OH^- \longrightarrow$$

$$Fe^{3+} + H_2O_2\ Fe^{2+} + OOH\bullet + H^+ \longrightarrow$$

2.9.2.2   Advanced Oxidation Methods

2.9.2.2.1   Photocatalysis

Photocatalysis is the process of accelerating a photochemical shift using a catalyst including TiO_2 or Fenton's reagent. Rutile TiO_2 is the most often used catalyst in all pharmaceutical photocatalytic investigations. Apart from TiO_2, ZnO_2 show higher catalytic activity for the degradation of chloramphenicol and sulfamethazine present in pharmaceutical effluent. Photocatalysis is the optimum technique for effluents with high COD effluents and for the complete conversion of extremely refractory organic pollutants to reach biological treatment levels. Photocatalytic reactions typically follow the Langmuir-Hinshelwood kinetic model, which can be simplified to pseudo-first-order or zero-order kinetics depending on the operating parameters (Gadipelly et al., 2014).

2.9.2.2.2 Photo-Fenton

Photo-Fenton reaction involves the introduction of UV to the Fenton system that promotes the reduction of Fe^{3+} to Fe^{2+}. H_2O_2/UV is mixed with iron salts in photo-Fenton to accelerate radical generation that can be effectively used for degrading pharmaceutical wastewater. Hydroxyl ions can be synthesized in two ways including excitation of $Fe(OH)^{2+}$ and direct photolysis of H_2O_2.

$$Fe\ (OH)^{2+} + h\nu \qquad Fe^{2+} + OH \cdot \qquad \longrightarrow$$

$$H_2O_2 + h\nu\, 2\ OH \cdot \qquad \longrightarrow$$

Photo-Fenton can be used as an alternative tertiary wastewater treatment process in pharmaceutical plants. Recent studies have reported the remarkable potential of photo-Fenton in degrading emerging pharmaceuticals of concern such as ibuprofen, carbamazepine, diclofenac, and erythromycin. Moreover, various sewage contaminants also exhibited considerable degradation with percentage degradation ranging from 70% to 99.5% upon treatment using the photo-Fenton reaction. However, one of the downsides of photo-Fenton involves the high consumption of catalysts and the production of iron sludge (Eniola et al., 2022).

2.9.3 BIOLOGICAL TREATMENT STRATEGIES

The elimination of PCs using biological degradation plays a significant role in protecting the environment from the hazardous impacts of PCs. This remediation approach has numerous benefits, including moderate working conditions, and an uncomplicated and cost-effective operating model. The elimination process occurs by using PCs as a substrate during microbial growth, which subsequently leads to the conversion and destruction of pharmaceutical waste. Various microbes are also recognized to work synergistically for removing PCs. Microorganisms need a primary substrate as a growth source in the biodegradation process for removing pharmaceutical wastes. Furthermore, certain microorganisms are able to use pharmaceutical pollutants as a sole carbon and energy source, which eventually leads to process' cost reduction. In the biological system, the investigations exploring the elimination of pharmaceutical wastes focus on parameters such as the effectiveness of treatment, sorption method, and degradation capacity. The performance of the biological removal technique is dependent on seasonal changes. Moreover, in some cases, the formation of secondary contaminants with sludge is also reported after the biodegradation of these contaminants, which is a matter of concern in the scientific world (Simon et al., 2021). Therefore, it is important to design a treatment method capable of overcoming challenges associated with the formation of secondary contaminants. Different biological treatment strategies for pharmaceutical wastewater are as follows.

2.9.3.1   Activated Sludge Process

The activated sludge process (ASP) is a widely employed procedure for the biological treatment of wastewater containing a variety of pharmaceuticals. The conventional activated sludge (CAS) treatment is a cost-efficient technology that is primarily determined by two parameters including hydraulic retention period (HRT) and temperature. Other parameters that impact the efficacy of the ASP include the presence of organic matter, pH, BOD, COD, and the presence of non-biodegradable waste. The biological treatment removes pharmaceutical compounds by a combination of adsorption, volatilization, and biodegradation. However, the elimination of pharmaceuticals by adsorption and volatilization by activated sludge is considerably less as volatilization only contributes to remove less than 10% of most of the toxic pharmaceuticals. Moreover, the physicochemical features of the targeted compounds and environmental parameters including temperature and pH strongly affect the adsorption of pharmaceutical pollutants to activated sludge. Generally, biodegradation is the primary process for the efficient removal of PCs by activated sludge but it is not always successful because of the limited abundance of degraders in the natural environment. This constraint can be overcome by biological acclimation and bio-augmentation that enhance the aggregates of pure or mixed culture for the effective degradation of contaminants in the biological treatment process (Wang and Wang, 2016). Studies have shown the efficacy of the CAS method to remove as well as bio-transform antibiotics such as sulfamethoxazole, erythromycin, norfloxacin, ciprofloxacin, and roxithromycin. Furthermore, cefaclor has been reported to have a removal efficiency of 98% while spiramycin exhibits inconsiderable removal upon treatment with activated sludge. However, the removal efficiency of CAS has been found to be lesser than the membrane bioreactors (MBRs) approach as it shows the significant removal of azithromycin only when compared to the MBRs treatment process (Simon et al., 2021; Verlicchi and Zambello, 2014).

2.9.3.2   Membrane Bioreactor Technology

Membrane bioreactor (MBR) technology integrates both biological and physical processes (Membrane filtration) and is viewed as a major advancement in the microbiological treatment of toxic effluents. Following the removal of biodegradable materials in the biological process, the wastewater is further filtered by removing solutes via membrane filtration to obtain the effluent of the desired purity. When compared to traditional treatment strategies, membrane bioreactors have improved sludge retention, higher efficiency, and the ability to resist varying loads of contaminants (Tambosi et al., 2010; Eniola et al., 2022).

MBRs are generally efficient in degrading hydrophobic and readily biotransformable agents as they are easily absorbed in MBRs but are less effective at degrading hydrophilic and ecologically persisting drugs. The sorption capacity of a compound is also affected by the physicochemical properties of the membrane used. Hydrophobic membranes (polyethersulfone membranes) can retain lipophilic pharmacological compounds, charged membranes can retain pharmaceuticals via electrostatic interactions, whereas hydrophilic or neutrally charged membranes cannot considerably retain the pharmaceuticals. Moreover, various factors that influence the degradation efficacy of pharmaceuticals by MBRs include operational parameters (temperature, pH, conductivity, biomass concentration, sludge retention time, redox conditions), characteristics of the membrane, wastewater constituents, and pharmaceutical compounds' molecular weight, complexity and hydrophobicity/hydrophilicity (Nguyen et al., 2021).

Studies have revealed the promising potential of membrane bioreactor technology (hybrid technique) to remove pharmaceuticals (antifungal, anti-histamines, anti-inflammatory, hormones, and lipid regulator drugs) from effluent. The hybrid system (MBRs) is reported to have a removal efficacy of more than 84% for pharmaceuticals and a COD removal efficiency of 98% because a hybrid system uses both biotic and abiotic techniques in contrast to a single bioreactor that is without membrane filter. Some of the pharmaceutical compounds that show considerable transformation using MBRs include ibuprofen, metformin, naproxen, acetaminophen, paraxanthine, 2-hydroxyibuprofen, atorvastatin, caffeine, and cotinine. Besides, pharmaceuticals like carbamazepine, trimethoprim, thiabendazole, erythromycin, clarithromycin, and meprobamate are not significantly removed by the MBR process. Thus, it has been suggested to investigate the performance of MBR for longer SRTs and HRTs for the degradation of pharmaceuticals exhibiting insufficient degradation by sorption and biological processes (Kim et al., 2014; Simon et al., 2021).

2.9.3.3 Constructed Wetlands

Constructed Wetlands (CWs), a green technology, are miniature semi-aquatic ecosystems that play a pivotal role as an alternative strategy for the treatment of industrial, municipal, and agricultural wastewater. These wetlands are constructed artificially in man-made systems and are often long, narrow channels or trenches. Compared to conventional treatment systems, CWs offer low energy, and low operational requirements but are land-intensive thus requiring enormous expanses (Krzemiński and Popowska, 2020). Pollutant degradation in constructed wetlands occurs through a combination of three major processes such as physical, chemical, and biological degradation which involve various techniques including volatilization, sedimentation, sorption, accumulation, phytodegradation, plant uptake, and microbial biodegradation (Zhang et al., 2023). Depending upon the removal efficacy of pollutants in a constructed wetland, the pharmaceuticals are classified as readily, moderately, low, and hardly removed. Readily removed pharmaceuticals have a removal efficacy of 70% and these include acetaminophen, sulfadiazine, sulfadimethoxine, salicylic acid, sulfapyridine, sulfamethoxazole, sulfapyridine, sulfapyridine, tetracycline, trimethoprim, atenolol, caffeine, and furosemide. Besides, moderately removed pharmaceuticals have a removal efficiency ranging between 50% and 70% and these antibiotics include ibuprofen, doxycycline, gemfibrozil, and naproxen. Moreover, low-removed pharmaceuticals have a mean removal efficacy between 20% and 50%, and these drugs majorly include triclosan, clofibric acid, carbamazepine, amoxicillin, diclofenac, and ketoprofen (Li et al., 2014).

Various factors that affect the degradation of pharmaceutical pollutants by constructed wetlands include soil/substrate matrix, design configuration, and the plant species used. Sub-surface flow constructed wetlands have been reported to show higher efficiency to remove biodegradable substances because this setting promotes the interaction between wastewater, plants, soil, and microorganisms. Apart from this, the surface flow constructed wetlands exhibit improved performance for compounds prone to photo-degradation. Photo-degradation can be carried out either through the direct exposure of water to sunlight or by the formation of photo-generated radicals (other than hydroxyl) in the indirect photolysis process.

2.9.3.4 Anaerobic Treatment

Anaerobic treatment can be accomplished by employing different reactor systems including fluidized bed reactors, continuous stirred tank reactors (anaerobic digestion), and up-flow anaerobic sludge reactors. In recent times the use of anaerobic hybrid reactors that integrates both suspended and attached growth systems has gained significant attention. The advantages of anaerobic treatment above aerobic procedures include the capability to manage high concentrations of wastewater while consuming less energy, producing less sludge, economically recovering bio-methane as a useful energy source, and having a lower operating cost (Sponza and Çelebi, 2012; Gadipelly et al., 2014).

Recent investigations have shown the removal efficiency of up-flow anaerobic batch reactors (USAR) that have the potential to remove a substantial load of pharmaceutical pollutants from pharmaceutical wastewater with COD and BOD removal efficiencies ranging from 65%–75% to 80%–94% respectively at the higher temperature of 55°C. However, combining catalytic wet air oxidation with anaerobic biological oxidation has been shown to enhance COD removal by more than 94.66% accompanied by the biodegradation of the organic content present in the effluent (Gadipelly et al., 2014). Moreover, the coupling of a continuous stirred tank reactor (CSTR) with an anaerobic multi-chamber bed reactor (AMCBR) is reported to be effective in removing oxytetracycline while being ineffective in eliminating kitasamycin and colistin sulfate from antibiotic-containing wastewater (Tang et al., 2011; Gadipelly et al., 2014).

2.10 REGULATORY GUIDELINES FOR PHARMACEUTICAL EFFLUENTS

Global pharmaceutical industries have shown tremendous growth over the past few decades motivated by the rise in demand for pharmaceuticals due to increasing lifespan, growing economies, abrupt climate change, and evolution of clinical practices. It has been estimated that the global pharmaceutical market is expanding at the rate of 6.5% per annum but this progress is beclouded by the unregulated disposal of pharmaceuticals and their metabolites in the environment leading to unintended deleterious impacts on aquatic life and the environment. It has been estimated by the OECD report (2019) that 88% of human pharmaceutical products lack comprehensive data on environmental toxicity. While pharmaceuticals are rigorously monitored for patient safety and efficacy, the adverse effects that they may induce on the natural environment have not been adequately investigated (OECD, 2019).

Pharmaceutical pollution management necessitates the need for novel wastewater treatment technologies and behavioral adjustments in industry, the health care system, and society at large. Various organizations around the globe have established different regulatory guidelines for managing active pharmaceuticals in the environment. In the US, EPA has established effluent limitation guidelines under the Clean Water Act for regulating the uncontrolled discharge of wastewater from different pharmaceutical manufacturing units. These regulations are devised to establish the "best practicable control technology (BPT)", "best available technology (BAT)", "best conventional pollutant control technology (BCT)", "new source performance standards (PSES)", and "pretreatment standards for existing and new sources (PSNS)" for certain pollutant being manufactured and discharged directly or indirectly into the fresh waterbodies by different pharmaceutical facilities. Apart from this, OECD recommends that all the concerned governmental departments, stakeholders, and local authorities should collaborate in designing and promulgating a mix of source-oriented, user-directed, and end-of-pipe measures (OECD, 2019; EPA, 1983). Various policies that institute, encourage and incentivize the measures for preventing the expulsion of toxic pharmaceuticals into waterbodies and their harmful impacts on the aquatic ecosystem are as follows.

i. Industries should adopt clean manufacturing practices, environmental management systems, and the 3R strategy (Reduce, Reuse and Recycle) to avoid, eliminate, and reduce the production of hazardous PCs.

ii. Pharmaceutical industries should be encouraged to employ sustainable approaches and green technology, such as Best Available Techniques and Best Environmental Practices (BAT & BEP), to promote greener and cleaner design and manufacturing.

iii. Pharmaceutical units should take measures to minimize waste production by investing in research, design, innovation, and the development of novel processes that aid in reducing and eliminating the production of toxic pharmaceutical waste.

iv. Industries should ensure the use of alternate green chemicals that are eco-friendly, biodegradable, and have low intrinsic toxicity and benign design.

v. Government should make efforts to upgrade and strengthen the existing national and private facilities for testing hazardous waste and new testing facilities should be established and accredited as per the requirement.

vi. Pharmaceutical industries should make technical human resources available for supervision, assistance, and monitoring of waste management operations in their facilities.

vii. Industries must develop an inventive and targeted risk assessment strategy to determine the risks associated with the disposal of active pharmaceutical ingredients into freshwater bodies and their impact on the ecosystem.

viii. The government and stakeholders should try to ensure the transparency, robustness, and consistency of the Environmental Risk Assessment (ERA).

ix. Academic and research institutes should collaborate with industries to reduce the knowledge gap and devise guidelines and strategies for the management of hazardous waste generated in industries.

x. Governmental departments and local authorities should try to educate and engage the public to raise awareness and control the perceived and actual risks associated with the unregulated disposal of pharmaceuticals.

  xi. In order to incentivize emission reduction, pharmaceutical companies should pay taxes for discharging wastewater to waterbodies.

 xii. Government should provide financial assistance to encourage operators to invest in advanced wastewater treatment or to facilitate research on improved wastewater treatment.

xiii. Government must provide assistance to industries in the form of information, guidance, and consultancy on efficient wastewater treatment or solid waste management.

xiv. The establishment of Environmental Quality Norms (EQNs) and water quality standards should be ensured for identifying hazardous pollutants in waterbodies and if they are detected above safe levels, actions should be taken to prevent the water bodies from the harmful effects of these contaminants.

 xv. Pharmaceutical manufacturing plants should be issued effluent discharge permits with obligations to protect drinking water sources and freshwater habitats. Noncompliance may result in fines or the revocation of operating licenses.

2.11   SUMMARY AND CONCLUSIONS

Pharmaceuticals released into wastewater have emerged as a pressing environmental issue, stemming from diverse sources like unused medications, industrial discharge, and human waste. Their infiltration into water bodies poses multifaceted risks: disrupting aquatic ecosystems, accumulating in the food chain, and potentially impacting human health through prolonged exposure or fostering antibiotic-resistant bacteria. Addressing these concerns, multiple wastewater treatment technologies have been developed. Physical methods like screening and membrane separation, chemical processes such as oxidation, and biological techniques utilizing microbial activity or natural processes like constructed wetlands aim to remove pharmaceuticals. The choice of treatment technology depends on factors such as the specific pharmaceuticals present, their concentrations, and the overall characteristics of the wastewater. Effective treatment not only safeguards the environment but also helps mitigate potential risks to human health associated with the presence of pharmaceuticals in wastewater.

In the last few decades, pharmaceutical pollutants have emerged as major contaminants of concern in wastewater. The ubiquitous occurrence of these contaminants in the aquatic environment has gained much attention because of their adverse impacts on humans, animals, and the ecosystem. Pharmaceuticals enter the environment via different means with pharmaceutical manufacturing units being one of the most important routes. Multiple research projects are underway to detect pharmaceutical pollutants in wastewater and develop an appropriate technique for their complete removal. The present study describes different pharmaceuticals, their adverse effects on human health and the environment, various treatment methods (including physical, chemical, and biological), and regulatory guidelines for their management. An appropriate method for removing pharmaceutical pollutants from wastewater can be chosen depending on the available space, procurable capital, and pollutant types.

Most of the documented treatment processes are still unable to carry out the complete removal of PCs. So, the improvement of present methods for pharmaceutical wastewater treatment that favor environmental protection should be studied further, as this will not only support the elimination of pharmaceutical pollutants from wastewater but it will also assist in the maintenance of the ecosystem.

REFERENCES

Abdallah, M. A.-E., Nguyen, K.-H., Ebele, A. J., Atia, N. N., Ali, H. R. H. & Harrad, S. 2019. A single run, rapid polarity switching method for determination of 30 pharmaceuticals and personal care products in waste water using Q-Exactive Orbitrap high resolution accurate mass spectrometry. *Journal of Chromatography A*, 1588, 68–76.

Ahmed, F., Tscharke, B., O'brien, J. W., Thompson, J., Zheng, Q., Mueller, J. F. & Thomas, K. V. 2021a. Quantification of selected analgesics and their metabolites in influent wastewater by liquid chromatography tandem mass spectrometry. *Talanta*, 234, 122627.

Ahmed, S., Mofijur, M., Nuzhat, S., Chowdhury, A. T., Rafa, N., Uddin, M. A., Inayat, A., Mahlia, T., Ong, H. C. & Chia, W. Y. 2021b. Recent developments in physical, biological, chemical, and hybrid treatment techniques for removing emerging contaminants from wastewater. *Journal of Hazardous Materials*, 416, 125912.

Anis, S. F., Hashaikeh, R. & Hilal, N. 2019. Microfiltration membrane processes: A review of research trends over the past decade. *Journal of Water Process Engineering*, 32, 100941.

Ba, S., Haroune, L., Soumano, L., Bellenger, J.-P., Jones, J. P. & Cabana, H. 2018. A hybrid bioreactor based on insolubilized tyrosinase and laccase catalysis and microfiltration membrane remove pharmaceuticals from wastewater. *Chemosphere*, 201, 749–755.

Bhargava, A. 2016. Physico-chemical waste water treatment technologies: An overview. *Int J Sci Res Educ*, 4, 5308–5319.

Bhushan, S., Rana, M. S., Raychaudhuri, S., Simsek, H. & Prajapati, S. K. 2020. Algae-and bacteria-driven technologies for pharmaceutical remediation in wastewater. *Removal of Toxic Pollutants through Microbiological and Tertiary Treatment*. Elsevier. https://doi.org/10.1016/B978-0-12-821014-7.00015-0.

Bilal, M., Mehmood, S., Rasheed, T. & Iqbal, H. M. 2020. Antibiotics traces in the aquatic environment: Persistence and adverse environmental impact. *Current Opinion in Environmental Science & Health*, 13, 68–74.

Castanon, J. 2007. History of the use of antibiotic as growth promoters in European poultry feeds. *Poultry Science*, 86, 2466–2471.

Chauhan, J. S. & Kumar, S. 2020. Wastewater ferti-irrigation: An eco-technology for sustainable agriculture. *Sustainable Water Resources Management*, 6, 1–11.

Chen, J., Ma, X., Gai, Z., Li, W. & Dong, Q. 2014. Acute rapamycin (RAP) exposure induce developmental, neurobehavioral and immunal toxicities in embryonic zebrafish. *Environmental Chemistry*, 33, 556–561.

Cho, H.-H., Huang, H. & Schwab, K. 2011. Effects of solution chemistry on the adsorption of ibuprofen and triclosan onto carbon nanotubes. *Langmuir*, 27, 12960–12967.

Chu, W., Fang, C., Deng, Y. & Xu, Z. 2020. Intensified disinfection amid COVID-19 pandemic poses potential risks to water quality and safety. *Environmental Science & Technology*, 55, 4084–4086.

Crini, G. & Lichtfouse, E. 2019. Advantages and disadvantages of techniques used for wastewater treatment. *Environmental Chemistry Letters*, 17, 145–155.

Cuerda-Correa, E. M., Alexandre-Franco, M. F. & Fernández-González, C. 2019. Advanced oxidation processes for the removal of antibiotics from water: An overview. *Water*, 12, 102.

Cycon, M., Mrozik, A. & Piotrowska-Seget, Z. 2019. Antibiotics in the soil environment-degradation and their impact on microbial activity and diversity. *Front Microbiol* 10, 338.

Deegan, A., Shaik, B., Nolan, K., Urell, K., Oelgemöller, M., Tobin, J. & Morrissey, A. 2011. Treatment options for wastewater effluents from pharmaceutical companies. *International Journal of Environmental Science & Technology*, 8, 649–666.

Ebele, A. J., Abdallah, M. A.-E. & Harrad, S. 2017. Pharmaceuticals and personal care products (PPCPs) in the freshwater aquatic environment. *Emerging Contaminants*, 3, 1–16.

Eniola, J. O., Kumar, R., Barakat, M. & Rashid, J. 2022. A review on conventional and advanced hybrid technologies for pharmaceutical wastewater treatment. *Journal of Cleaner Production*, 356, 131826.

EPA 1983. Pharmaceutical Manufacturing Point Source Category Effluent Limitations Guidelines, Pretreatment Standards, and New Source Performance Standards. Vol. 48, No. 209.

Escudero, J., Muñoz, J., Morera-Herreras, T., Hernandez, R., Medrano, J., Domingo-Echaburu, S., Barceló, D., Orive, G. & Lertxundi, U. 2021. Antipsychotics as environmental pollutants: An underrated threat? *Science of the Total Environment*, 769, 144634.

Esfahani, N. R., Mobarekeh, M. N. & Hoodaji, M. 2018. Effect of grit chamber configuration on particle removal: Using response surface method. *Journal of Membrane and Separation Technology*, 7, 12–16.

Farzaneh, H., Loganathan, K., Saththasivam, J. & Mckay, G. 2020. Ozone and ozone/hydrogen peroxide treatment to remove gemfibrozil and ibuprofen from treated sewage effluent: Factors influencing bromate formation. *Emerging Contaminants*, 6, 225–234.

Feng, L., Van Hullebusch, E. D., Rodrigo, M. A., Esposito, G. & Oturan, M. A. 2013. Removal of residual anti-inflammatory and analgesic pharmaceuticals from aqueous systems by electrochemical advanced oxidation processes. A review. *Chemical Engineering Journal*, 228, 944–964.

Gadipelly, C., Pérez-González, A., Yadav, G. D., Ortiz, I., Ibáñez, R., Rathod, V. K. & Marathe, K. V. 2014. Pharmaceutical industry wastewater: Review of the technologies for water treatment and reuse. *Industrial & Engineering Chemistry Research*, 53, 11571–11592.

Ghazal, H., Koumaki, E., Hoslett, J., Malamis, S., Katsou, E., Barcelo, D. & Jouhara, H. 2022. Insights into current physical, chemical and hybrid technologies used for the treatment of wastewater contaminated with pharmaceuticals. *Journal of Cleaner Production*, 361, 132079.

Hassan, S. S., Abdel-Shafy, H. I. & Mansour, M. S. 2019. Removal of pharmaceutical compounds from urine via chemical coagulation by green synthesized ZnO-nanoparticles followed by microfiltration for safe reuse. *Arabian Journal of Chemistry*, 12, 4074–4083.

Kayode-Afolayan, S. D., Ahuekwe, E. F. & Nwinyi, O. C. 2022. Impacts of pharmaceutical effluents on aquatic ecosystems. *Scientific African*, 17, e01288.

Kim, M., Guerra, P., Shah, A., Parsa, M., Alaee, M. & Smyth, S. 2014. Removal of pharmaceuticals and personal care products in a membrane bioreactor wastewater treatment plant. *Water Science and Technology*, 69, 2221–2229.

Kotwani, A., Joshi, J. & Kaloni, D. 2021. Pharmaceutical effluent: A critical link in the interconnected ecosystem promoting antimicrobial resistance. *Environmental Science and Pollution Research*, 28, 32111–32124.

Krzemiński, P. & Popowska, M. 2020. Treatment technologies for removal of antibiotics, antibiotic resistance bacteria and antibiotic-resistant genes. *Antibiotics and Antimicrobial Resistance Genes*. Springer. https://doi.org/10.1007/978-3-030-40422-2_19.

Kumar, M., Jaiswal, S., Sodhi, K. K., Shree, P., Singh, D. K., Agrawal, P. K. & Shukla, P. 2019. Antibiotics bioremediation: Perspectives on its ecotoxicity and resistance. *Environment International*, 124, 448–461.

Larsson, D. J. 2014. Antibiotics in the environment. *Upsala Journal of Medical Sciences*, 119, 108–112.

Li, Y., Zhu, G., Ng, W. J. & Tan, S. K. 2014. A review on removing pharmaceutical contaminants from wastewater by constructed wetlands: Design, performance and mechanism. *Science of the Total Environment*, 468, 908–932.

Lima, E. C. 2018. Removal of emerging contaminants from the environment by adsorption. *Ecotoxicology and Environmental Safety*, 150, 1–17.

Liu, F.-F., Zhao, J., Wang, S., Du, P. & Xing, B. 2014. Effects of solution chemistry on adsorption of selected pharmaceuticals and personal care products (PPCPs) by graphenes and carbon nanotubes. *Environmental Science & Technology*, 48, 13197–13206.

Łukaszewicz, P., Maszkowska, J., Mulkiewicz, E., Kumirska, J., Stepnowski, P. & Caban, M. 2016. Impact of veterinary pharmaceuticals on the agricultural environment: A re-inspection. *Reviews of Environmental Contamination and Toxicology*, 243, 89–148.

Mailler, R., Gasperi, J., Coquet, Y., Deshayes, S., Zedek, S., Cren-Olivé, C., Cartiser, N., Eudes, V., Bressy, A. & Caupos, E. 2015. Study of a large scale powdered activated carbon pilot: Removals of a wide range of emerging and priority micropollutants from wastewater treatment plant effluents. *Water Research*, 72, 315–330.

Martínez, J. L. 2017. Effect of antibiotics on bacterial populations: A multi-hierachical selection process. *F1000Research*, 6, 51.

Monapathi, M. E., Oguegbulu, J. C., Adogo, L., Klink, M., Okoli, B., Mtunzi, F. & Modise, J. S. 2021. Pharmaceutical pollution: Azole antifungal drugs and resistance of opportunistic pathogenic yeasts in wastewater and environmental water. *Applied and Environmental Soil Science*, 2021. doi: 10.1155/2021/9985398.

Naidoo, S. & Olaniran, A. O. 2014. Treated wastewater effluent as a source of microbial pollution of surface water resources. *International Journal of Environmental Research and Public Health*, 11, 249–270.

Nam, S.-W., Choi, D.-J., Kim, S.-K., Her, N. & Zoh, K.-D. 2014. Adsorption characteristics of selected hydrophilic and hydrophobic micropollutants in water using activated carbon. *Journal of Hazardous Materials*, 270, 144–152.

Nguyen, P., Carvalho, G., Reis, M. A. & Oehmen, A. 2021. A review of the biotransformations of priority pharmaceuticals in biological wastewater treatment processes. *Water Research*, 188, 116446.

Nunes, B., Antunes, S., Gomes, R., Campos, J., Braga, M., Ramos, A. & Correia, A. 2015. Acute effects of tetracycline exposure in the freshwater fish Gambusia holbrooki: Antioxidant effects, neurotoxicity and histological alterations. *Archives of Environmental Contamination and Toxicology*, 68, 371–381.

Obotey Ezugbe, E. & Rathilal, S. 2020. Membrane technologies in wastewater treatment: A review. *Membranes*, 10, 89.

OECD 2019. Pharmaceutical Residues in Freshwater: Hazards and Policy Responses. *OECD Studies on Water*.

Parveen, N., Chowdhury, S. & Goel, S. 2022. Environmental impacts of the widespread use of chlorine-based disinfectants during the COVID-19 pandemic. *Environmental Science and Pollution Research*, 29, 1–19.

Pickering, K. T. & Hiscott, R. N. 2015. *Deep Marine Systems: Processes, Deposits, Environments, Tectonics and Sedimentation*. John Wiley & Sons, Hoboken, NJ.

Ramin, E., Wágner, D. S., Yde, L., Binning, P. J., Rasmussen, M. R., Mikkelsen, P. S. & Plósz, B. G. 2014. A new settling velocity model to describe secondary sedimentation. *Water Research*, 66, 447–458.

Ricceri, F., Giagnorio, M., Zodrow, K. R. & Tiraferri, A. 2021. Organic fouling in forward osmosis: Governing factors and a direct comparison with membrane filtration driven by hydraulic pressure. *Journal of Membrane Science*, 619, 118759.

Rosenblat, J. D. & Mcintyre, R. S. 2020. Pharmacological treatment of major depressive disorder. *Major Depressive Disorder*. Elsevier. https://doi.org/10.1016/B978-0-323-58131-8.00008-2.

Samal, K., Mahapatra, S. & Ali, M. H. 2022. Pharmaceutical wastewater as Emerging Contaminants (EC): Treatment technologies, impact on environment and human health. *Energy Nexus*, 6, 100076.

Samsami, S., Mohamadizaniani, M., Sarrafzadeh, M.-H., Rene, E. R. & Firoozbahr, M. 2020. Recent advances in the treatment of dye-containing wastewater from textile industries: Overview and perspectives. *Process Safety and Environmental Protection*, 143, 138–163.

Shahtalebi, A., Sarrafzadeh, M. & Montazer, R. M. 2011. Application of nanofiltration membrane in the separation of amoxicillin from pharmaceutical wastewater. *Iranian Journal of Environmental Health Science & Engineering*, 8, 109–116.

Shakoor, S., Hasan, Z. & Hasan, R. 2020. Epidemiological, ecological, and public health effects of antibiotics and AMR/ARGs. *Antibiotics and Antimicrobial Resistance Genes*. Springer. https://doi.org/10.1007/978-3-030-40422-2_12.

Shang, A. H., Ye, J., Chen, D. H., Lu, X. X., Lu, H. D., Liu, C. N. & Wang, L. M. 2015. Physiological effects of tetracycline antibiotic pollutants on non-target aquatic Microcystis aeruginosa. *Journal of Environmental Science and Health, Part B*, 50, 809–818.

Simon, M., Kumar, A. & Garg, A. 2021. Biological treatment of pharmaceuticals and personal care products (PPCPs) before discharging to environment. *Fate and Transport of Subsurface Pollutants*. Springer. https://doi.org/10.1007/978-981-15-6564-9_14.

Sponza, D. T. & Çelebi, H. 2012. Removal of oxytetracycline (OTC) in a synthetic pharmaceutical wastewater by sequential anaerobic multichamber bed reactor (AMCBR)/completely stirred tank reactor (CSTR) system: Biodegradation and inhibition kinetics. *Journal of Chemical Technology & Biotechnology*, 87, 961–975.

Tambosi, J. L., Yamanaka, L. Y., José, H. J., Moreira, R. D. F. P. M. & Schröder, H. F. 2010. Recent research data on the removal of pharmaceuticals from sewage treatment plants (STP). *Química Nova*, 33, 411–420.

Tang, C.-J., Zheng, P., Chen, T.-T., Zhang, J.-Q., Mahmood, Q., Ding, S., Chen, X.-G., Chen, J.-W. & Wu, D.-T. 2011. Enhanced nitrogen removal from pharmaceutical wastewater using SBA-ANAMMOX process. *Water Research*, 45, 201–210.

Thapa, B. S., Pandit, S., Patwardhan, S. B., Tripathi, S., Mathuriya, A. S., Gupta, P. K., Lal, R. B. & Tusher, T. R. 2022. Application of microbial fuel cell (MFC) for pharmaceutical wastewater treatment: An overview and future perspectives. *Sustainability*, 14, 8379.

Tufail, A., Price, W. E., Mohseni, M., Pramanik, B. K. & Hai, F. I. 2021. A critical review of advanced oxidation processes for emerging trace organic contaminant degradation: Mechanisms, factors, degradation products, and effluent toxicity. *Journal of Water Process Engineering*, 40, 101778.

Verlicchi, P., Al Aukidy, M. & Zambello, E. 2015. What have we learned from worldwide experiences on the management and treatment of hospital effluent?-An overview and a discussion on perspectives. *Science of the Total Environment*, 514, 467–491.

Verlicchi, P. & Zambello, E. 2014. How efficient are constructed wetlands in removing pharmaceuticals from untreated and treated urban wastewaters? A review. *Science of the Total Environment*, 470, 1281–1306.

Wang, J. & Chu, L. 2016. Irradiation treatment of pharmaceutical and personal care products (PPCPs) in water and wastewater: An overview. *Radiation Physics and Chemistry*, 125, 56–64.

Wang, J. & Wang, S. 2016. Removal of pharmaceuticals and personal care products (PPCPs) from wastewater: A review. *Journal of Environmental Management*, 182, 620–640.

WHO 2012. Pharmaceuticals in drinking-water.

Wilkinson, J., Boxall, A., Kolpin, D., Leung, K., Lai, R., Wong, D., Ntchantcho, R., Pizarro, J., Mart, J. & Echeverr, S. 2022. Pharmaceutical pollution of the world's rivers. *Proceedings of the National Academy of Sciences of the United States of America*, 119, 1–10.

Yang, G. C. & Tang, P.-L. 2016. Removal of phthalates and pharmaceuticals from municipal wastewater by graphene adsorption process. *Water Science and Technology*, 73, 2268–2274.

Yu, M., Wang, J., Tang, L., Feng, C., Liu, H., Zhang, H., Peng, B., Chen, Z. & Xie, Q. 2020. Intimate coupling of photocatalysis and biodegradation for wastewater treatment: Mechanisms, recent advances and environmental applications. *Water Research*, 175, 115673.

Zhang, H., Wang, X. C., Zheng, Y. & Dzakpasu, M. 2023. Removal of pharmaceutical active compounds in wastewater by constructed wetlands: Performance and mechanisms. *Journal of Environmental Management*, 325, 116478.

Zhang, S., Abbas, M., Rehman, M. U., Huang, Y., Zhou, R., Gong, S., Yang, H., Chen, S., Wang, M. & Cheng, A. 2020. Dissemination of antibiotic resistance genes (ARGs) via integrons in *Escherichia coli*: A risk to human health. *Environmental Pollution*, 266, 115260.

3 Toxic Metals Accumulation in Fungi for Wastewater Treatment Plant

Maitri Nandasana and Sougata Ghosh

3.1 INTRODUCTION

The poisoning of water bodies by heavy metals is considered a major concern on a global scale. Heavy metals like arsenic, chromium, copper, iron, lead, mercury, and nickel are released into the environment via industrial processes like mining, metal processing, iron-sheet cleaning, chemical manufacturing, plating, automobiles, fertilizer, pesticide, and petroleum [1,2]. These toxic heavy metals are considered a serious threat to both the environment and human health. Metals are very easily absorbed or taken up by organisms due to their high solubility in aquatic media. As metals are non-biodegradable, they increase higher risks to living things because of their high toxicity, persistence in nature, and prolonged half-lives after entering into the food chain[3,4]. Exposure to different heavy metals results in several illnesses, including impairment of kidney function, rheumatoid arthritis, nervous system disorders, joint problems, and damage to the developing brain in humans. Children coming in contact with toxic metals have a higher risk of developing cardiovascular diseases, impaired development, and reduced intelligence [5].

Although the issue is seen in both developed and developing nations, the latter are more affected due to a lack of a progressive workforce, innovative technologies, and low policy implementation. Numerous conventional techniques, such as evaporation, ion exchange, chemical precipitation, coagulation, filtration, membrane technology, redox, reverse osmosis, carbon adsorption, electrowinning, electrochemical treatment, and chelation, are already being used in the eradication of these contaminants from the aquatic environment [6–14]. These technologies have limitations in terms of complexity, cost-effectiveness, the possibility of forming secondary pollutants, and their ability to change the physical and chemical makeup of the environment. Moreover, metal pollution also results in the elimination of non-targeted beneficial fauna as well as microbial species, including nitrogen-fixing bacteria [15]. There is already a huge demand for the affordable and effective methods to eliminate heavy metals from the industrial effluent.

Nowadays, the development of different biological techniques resulted in an environment-friendly remediation to address heavy metal contamination. The enhanced

DOI: 10.1201/9781003368120-3

biological methods also showed better efficiency in the treatment of high-volume effluents (low biomass) along with the rapid turnover time. Bioaugmentation, bio-transformation, biosorption, biofilters, composting, biostimulation, and bioreactors, are some of the most well-known bioremediation processes [16]. In the biological method, different microbes like bacteria, fungi, and algae, are used as adsorbents for the elimination of heavy metals. Microorganisms can convert heavy metals from a toxic to a low toxic form. They also take part in the accumulation or adsorption of different heavy metals [17,18].

Compared to other biosorption agents, fungal biomasses have a significant amount of cell wall material, which exhibits outstanding metal-binding characteristics due to their significant metal tolerance and other adverse environmental conditions, including low pH [19,20]. Various species of fungi such as *Aspergillus, Fusarium, Penicillium, Rhizopus, Trichoderma*, etc. are reported for the removal of different heavy metals.

This chapter gives elaborated information about the elimination of various heavy metals from wastewater with the help of different fungal strains and also about the mechanisms involved for tolerance against heavy metals and removal along with the effects of different parameters like agitation rate, contact time, dosages of adsorbent, solution pH, initial metal concentration, and temperature on the overall metal removal efficiency.

3.2 CADMIUM

Various fungi can effectively remove the hazardous cadmium (Cd) metal which is associated with industries related to electronics, pigments, metallurgy, and mining. Kumar et al. [21] studied the cadmium (Cd) tolerance and its removal efficiency by the filamentous fungus *Fusarium solani*. The plate assay method was employed to study the fungal tolerance to heavy metals such as Cd, nickel (Ni), and lead (Pb). The solutions with different metal concentrations (1,000, 500, 200, and 100 mg/L) were treated with fungal mycelium grown on the potato dextrose agar (PDA) plates and incubated for 6 days at $28°C \pm 1°C$. The Cd removal potential of *F. solani* was tested by conducting batch experiments containing 100 mL Cd of different concentrations that were amended in potato dextrose broth (PDB) medium with pH 6. Further, 1 mL spore suspension (2.7×10^4 mL) of *F. solani* was inoculated in the reaction mixture followed by incubation under continuous shaking conditions at 120 rpm for 5 days at $28°C \pm 1°C$. After incubation, around 10 mL of the PDB medium was centrifuged at 8,000 rpm for 8 minutes. After centrifugation, the obtained supernatant was checked for the residual Cd concentration. The tolerance analysis suggested that *F. solani* was completely inhibited by Ni and Pb, although Cd-mediated inhibition was observed after 4 days. The tolerance profile was nearly similar for 1,000, 500, 200, and 100 mg/L concentrations of heavy metals. The highest tolerance was observed against Cd followed by Ni and Pb. The highest Cd removal efficiency of 89.8% was noted for *F. solani* at pH 10 which suggested the strong attraction between the cell surface (negative charge) and Cd ions (positive charge). After 96 hours of incubation, the effect of Cd concentration showed an alteration in removal percentage from 50 mg/L ($87.5\% \pm 4.2\%$) to 200 mg/L ($42.9\% \pm 2.7\%$) Cd concentration.

The highest removal of Cd was observed at 144 hours (92.4% ± 5.8%) followed by 120 hours (87.5% ± 4.2%) and the lowest at 24 hours of incubation time. The increase in the level of the oxidized glutathione (GSSG) and glutathione (GSH) provides evidence in support of the role of Cd ions in the generation of reactive oxygen species (ROS). The increase in Cd concentration (100 mg/L) was proportional to a significant increase in the amount of GSH (90.37%) and GSSG (317%). At 100 mg/L Cd concentration, catalase (CAT) showed the highest antioxidant activity around 15 ± 0.1 μmol H_2O_2/min/mg protein. The high volume of H_2O_2 around 9.97 ± 0.07 μmol/g FW was found at a Cd concentration of 100 mg/L. The comparison of Fourier transform infrared spectroscopy (FTIR) analysis peaks of the control PDB medium and fungal biomass grown in Cd amended PDB revealed the disappearance of peaks attributed to the stretching vibrations at 1,397.0 cm^{-1} (carboxylate anions), 1,740.0 cm^{-1} of C=O in the carboxylic group, and 1,450.0 cm^{-1} representing the C-H and N-H stretching of amide after Cd (50 mg/L) removal. It was concluded that the electrostatic interaction between carboxylate anion and Cd^{2+} ions as well as between lone pairs and Cd^{2+} ions indicated the biosorption of Cd. Two new peaks found in the presence of 50 and 120 mg/L Cd concentration were mainly assigned to the N-H stretching (1,413.0 cm^{-1}) and C-H stretching (1,415.8 cm^{-1}). The shifting of the peaks (50 mg/L and 120 mg/L Cd concentration) from 1,072.0 to 1,080.6 and 1,075.2 cm^{-1} revealed that the phosphate group was responsible for the Cd precipitations in the form of $Cd_3(PO_4)_3$. The scanning electron microscope (SEM) analysis was insufficient to identify the visible damages of the fungal biomass before and afterward Cd treatment. The atomic percentage of Cd after treatment was 0.13% (50 mg/L) and 0.63% (120 mg/L). The bio-precipitation of Cd in the form of cadmium phosphate was suggested from the increase in the phosphorous weight percentage from 1.07% (control) to 2% (50 mg/L), but it reduced on the treatment with 120 mg/L Cd. At 50 and 120 mg/L Cd concentration, the sulfur weight percent was 0.28%, 0.46%, and 0.53% in control suggesting the increase in cadmium sulfide (CdS) in the biomass. The X-ray diffraction (XRD) analysis revealed a strong 2θ peak at 28.4° in the Cd-treated biomass attributing mainly to the $Cd_3(PO_4)_2$ formation.

Alothman et al. [22] used the fungi *Penicillium chrysogenum* and *Aspergillus ustus* to develop a low-cost biosorbent for the removal of copper (Cu (II)), Cd (II), and Pb (II) from the wastewater. The mycelium was harvested from fungal isolates which grew in the Cazapec liquid media and were further incubated at 28°C for 5 days. The mycelium was obtained through filtration followed by drying in an oven. The liquid nitrogen was used to ground the dried mycelium. In 10 mL of 1 ppm Cd (II), Cu (II), and Pb (II), at pH 5, 6, and 7, 2 g percent of dry biomass was added. The treated bottles were further incubated for 20 minutes at room temperature under shaking conditions (100 rpm). The surface charge of the fungal isolates (*P. chrysogenum* and *A. ustus*) was negative at pH < 7. The ideal pH for the elimination of Cu and Cd was 5 and 6 while using *P. chrysogenum* and *A. ustus*, and for Pb (II) was 5 and 7, respectively. The highest biosorption of around 91% of Cd (II), 53% of Cu (II), and 56% of Pb (II) were noted in the presence of the *P. chrysogenum* while around 84% of Cd (II), 52% of Cu (II), and 42% of Pb (II) was obtained by the use of *A. ustus*. The electrostatic interaction, metal ion exchange with the biosorbents, and complex formation were all possible paths for heavy metal adsorption.

Yaghoubian et al. [23] investigated the elimination of Cd by using filamentous fungi *Piriformospora indica* and *Trichoderma* spp. The dilution plate method was used for the fungi isolation. The growth of *Trichoderma* spp. was attained using a sterile PDB medium in which the Cd concentrations were gradually increased from 0 to 500 mg/L as shown in Figure 3.1. The required amount of stock solution of $CdCl_2 \cdot 2.5\ H_2O$ was added to the growth medium in order to achieve the desired concentration of Cd. The fungus agar discs (4 mm) of *Trichoderma* isolates (7-day-old culture) and *P. indica* (10-day-old culture) were inoculated followed by incubation for 15 days at 28°C (only for *Trichoderma*) under continuous shaking conditions. It was observed that, the *Trichoderma* spp. (*Trichoderma lixii, T. harzianum, T. simmonsii, T. longibrachiatum,* and *T. citrinoviride*) had high tolerance against Cd toxicity up to 500 mg/L concentration of Cd. At 500 mg/L concentration of Cd, only a 48.4% reduction in the biomass of *T. simmonsii* was noted. The minimum inhibitory

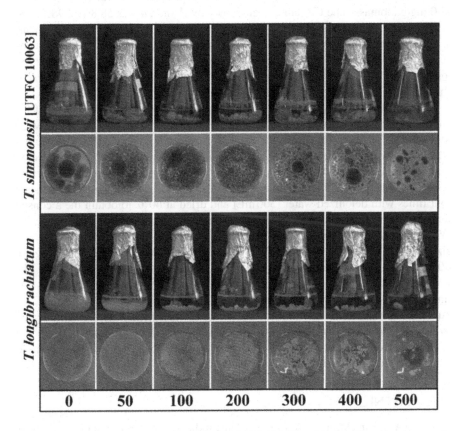

FIGURE 3.1 The response of *T. simmonsii* [UTFC 10063] and *T. longibrachiatum* growth to cadmium concentration gradients (mg/L) in the broth culture, 15 days after inoculation. Reprinted with permission from Yaghoubian, Y., Siadat, S.A., Moradi Telavat, M.R., Pirdashti, H., Yaghoubian, I., 2019. Bio-removal of cadmium from aqueous solutions by filamentous fungi: *Trichoderma* spp. and *Piriformospora indica. Environ. Sci. Pollut. Res.* 26, 7863–7872. Copyright © Springer-Verlag GmbH Germany, part of Springer Nature 2019.

concentration (MIC) of Cd which resulted in around 50% reduction of fungal biomass was 225 mg/L (*T. lixii*), 200 mg/L (*T. harziamum*), 325 and 500 mg/L (*T. simmonsii*), 175 mg/L (*T. citrinoviride*). A negative correlation was found between Cd uptake by *Trichoderma* sp. and fungal biomass production. The *Trichoderma* sp. was less able to accumulate Cd when its concentration increased from 0 to 500 mg/L. The highest Cd removal efficiency of 95.3% was recorded for *T. lixii* at 5 mg/L Cd concentration. At 10 and 500 mg/L Cd concentration, the *T. simmonsii* was able to remove around 91.7% and 31.2% Cd. Thus, *T. simmonsii* was the most potent organism for the accumulation of Cd among the rest of the *Trichoderma* spp. Compared to *Trichoderma* spp., *P. indica* showed lower tolerance against Cd. Only 30.4% and 7.5% biomass production were noted at 15 and 30 mg/L Cd concentration after 15 days of treatment whereas, at 200 mg/L Cd concentration no mycelial growth was detected. The highest Cd uptake of 105.4 mg/g biomass was noted for *P. indica* at 150 mg/L concentration of Cd while, in the *Trichoderma* spp. Cd uptake varied from 7.8 to 16.0 mg/g biomass. The Cd removal efficiency by *P. indica* was 58.8%, 2.98%, and 0.18% at 5, 50, and 200 mg/L Cd concentrations, respectively. The highest percentage of spore germination around 88.05% was noted for *P. indica* at 0 mg/L Cd concentration (control). The maximum spore formation of around 97.94% was observed for *T. simmonsii* at a 50 mg/L Cd concentration. The most resistant spores of *T. lixii* showed spore germination of 3.65% in 300 mg/L Cd concentration.

Rozman et al. [24] used waste fungal biomass (WFB) for the elimination of Cd and Pb ions. The WFB was attained after *Ganoderma lucidum* fruiting bodies were formed. The WFB was mostly made up of *G. lucidum* residue and the substrate (on which fungi were grown). The WFB was dried and pretreated in 300 mL of selected reagent (0.5 M NaOH, 0.5 M NaCl, and 10% (v/v) H_2O_2) for 1 hour under stirring condition (125 rpm) at room temperature (22°C ± 1°C). The biomass was washed four times with deionized water (300 mL) and dried at 60°C to obtain the a consistent mass. The field emission scanning electron microscopy (FE-SEM) images of WFB revealed a uniform surface with a rough texture and irregular pattern. The pore size of WFB was around 37.5 nm. The WFB particle size was found in the range of 40–100 μm with the average number of particles with a size of 73.7 ± 38.0 μm. The pretreated WFB achieved equilibrium in the Pb (II) adsorption in just 6 minutes and untreated WFB took around 18 minutes. Similarly, in the case of Cd adsorption, the equilibrium was achieved in 18 minutes (untreated WFB) and 6 minutes (pretreated WFB). The adsorption isotherm analysis revealed the Freundlich isotherm model ($R^2 > 0.9$) was best-fitted for the process of Cd (II) and Pb (II) on WFB, which indicates the heterogenous surface adsorption process.

3.3 ARSENIC

Arsenic (As) is an extremely hazardous element that can cause severe toxicity during prolonged exposure. Various fungi have the capability to remove As from the surroundings. Kumar and Dwivedi et al. [25] investigated Pb, Ni, Chromium (Cr), and arsenic (As) elimination from the wastewater by using *A. flavus* CR500 in the batch mode. The 100 mL of PDB medium containing Ni, Pb, and As separately at various concentrations (0, 10, 20, 50, and 100 mg/L) were prepared at pH 7. Around 1 mL fungal spore suspension (3×10^4 per mL) was inoculated in the flasks followed by 8 days of incubation at 80 rpm at $28 \pm 1°C$. After 8 days, the culture medium was subjected to 10 minutes of centrifugation at $4°C$ at 5,000 rpm and the resulting pellet was further kept at $60°C$ for drying for about 6 hours. The CR500 growth was inhibited severely by Cd and Cu at concentrations above 100 and 200 mg/L. The tolerance efficiency of CR500 was significant toward Ni, Pb, and As and therefore were selected for further analysis. During the single metal exposure experiment, a maximum removal of 99%, 97.7%, and 73% was noted at 10 mg/L concentration of As, Pb, and Ni metals, respectively after treatment with *A. flavus* CR500. The CR500 showed accumulation for Ni (4.91 mg/g), Pb (5.94 mg/g), and As (6.82 mg/g). At 100 mg/L concentration of Ni, As, and Pb, the surface absorption of efficiency of 0.12, 0.84, and 1.12 mg/g, respectively were noted. An increase in metal concentration adversely impacted the metal removal rate. At a metal concentration of 100 mg/L, biomass production was nearly unchanged in the case of Ni and As but decreased in the case of Pb (22.7%). The *A. flavus* CR500 had a high rate of biomass generation and outstanding multimetal removal potential from tannery wastewater (TWW) and simulated wastewater (SWW). The removal of 46.6%, 82.2%, 93.3%, and 97.5% of Ni, Cr, Pb, and As, respectively was obtained in SWW at 5 mg/L concentration of each metal.

Cardenas-Gonzalez et al. [26] studied As (V) elimination with the help of chemically modified biomasses of fungal mycelia. The biosorbents utilized in the removal of As (V) were *Cladosporium* sp., *Paecilomyces* sp., *A. fumigatus* I-II, *Mucor* sp-1 and 2, *A. flavus* III, IV, and V. At about $28°C$, the cultivation of the fungi was carried out in the liquid media comprising of 8 g/L thioglycolate under agitation and aeration. The cells formed after the completion of the incubation period (4–5 days) were centrifuged for 10 minutes ($700 \times g$) followed by washing with deionized water and a further 12 hours of drying at $80°C$. The pulverization of fungal biomass was performed followed by storage at $4°C$. The modification of the fungal biomass was carried out by iron oxide coating using 1 mL of 10 M NaOH, and 80 mL 2M $Fe(NO_3)_3 \cdot 9H_2O$. This mixture was further added to the porcelain pot containing 20 g of fungal biomass followed by 3 hours homogenization process at $80°C$. After raising the temperature to $110°C$ for 24 hours, the crushing method was employed in order to separate the coated biomass powder. At around 24 hours of incubation period and pH 6, the native biomass exhibited an As (V) removal efficiency of 10.1% (0.101 mg/L), whereas modified fungal biomass showed a high removal efficiency of 89% at similar conditions. This might be attributed to the dominant species ($H_2AsO_4^-$) at that pH which formed a complex with iron-coated biomass. The biomass coated with iron demonstrated the highest As (V) biosorption after incubating at 1 g/L biosorbent dose at pH 6 for 24 hours. The highest adsorption ability was observed at $30°C \pm 1°C$ whereas, a further increase in the temperature resulted in the reduction in absorption capacity of the iron-coated biomass of *Paecilomyces* sp. Interestingly, percentage absorption and ion concentration were inversely proportional. The increase in the *Paecilomyces* sp. biomass resulted in a higher number of biosorption sites which in turn aided in the higher As removal up to 99.3% at 5 g of fungal biomass within 24 hours. The modified fungal biomasses of *Paecilomyces* sp. and *A. flavus* IV, III, and V were found to have very effective As (V) removal ability equivalent to 97.1%, 92.3%, 90.3%, and 89%, respectively in the solution.

Hadiani et al. [27] evaluated As (V) and As (III) biosorption from the water resources using *Saccharomyces cerevisiae*. The yeast biomass was prepared in a sterilized liquid medium and the cell density used was 5×10^7 CFU/mL. With an increase in the pH from 3 to 7, the As biosorption increased that was maximum of up to 78% for As (III) and 19.7% for As (V) at 149 µg/L As concentration and pH 7. A higher As (V) and As (III) biosorption rates of 17% and 69%, respectively were noted at 49×10^7 CFU inoculum density and pH 6.3. The As concentration of 132.4 µg/L at pH 5 resulted in the highest biosorption efficiency of 66% for As (III)) and 16% for As (V) with an inoculum size of *S. cerevisiae* equivalent to 47.5×10^7 CFU. The biosorption rates increased with the increase in As (V) and As (III) ions in a range from 40 to 150 µg/L with the optimal concentration being 95 µg/L.

Govarthanan et al. [28] investigated the in-vitro bio-mineralization of Pb and As from aqueous solutions employing *Trichoderma* sp. The mineral precipitating media (MPM1) was prepared with 50–100 mM $CaCl_2$, pH (6–9), urea (100–500 mM), and 100 mg/L metal salts. The bio-mineralization experiment was performed by using the matrix containing soil (10 g) in 100 mg/kg Pb or As in MPM1, 75 mM $CaCl_2$ solution, 300 mM urea, and ten discs of *Trichoderma* culture. The solution was incubated for 14 days at 180 rpm and room temperature. The resulting product was used for extraction of metal after 48 hours of drying at 60°C. The maximum metal removal efficiencies of 59% (As) and 68% (Pb) and maximum urease production of 920 U/mL were noted at 75 mM $CaCl_2$, 300 mM urea, and pH 9. The elimination of metal was speculated to be directly correlated to the urease production. The mechanism behind the removal of Pb and As was evaluated using FTIR analysis. The alkanes, amide, and carboxyl groups were found to fulfill the role of preliminary molecules required in the successive precipitation of Pb and As.

3.4 MERCURY

Mercury (Hg) is an extremely hazardous metal that can be removed by various fungal isolates. Acosta-Rodriguez et al. [29] employed the resistant fungal strain *A. niger* in the elimination of different heavy metals from an aqueous solution. The fungal biomass was produced in 600 mL thioglycolate broth (8 g/L) after inoculating 1×10^6 spores/mL followed by incubation under shaking conditions (100 rpm) at 28°C. The cells were filtered after 5 days of incubation followed by two times washing and 12 hours of drying at 80°C. The biomass formed was further crushed and stored at room temperature. About 20 g of biomass powder was added to the mixture of 1 mL of 10 M NaOH and 80 mL of 2 M Fe $(NO_3)_3.9H_2O$ in a porcelain pot followed by homogenization and keeping it for 3 hours at 80°C to produce the iron oxide-coated biomass. The temperature was kept constant for 24 hours after being raised to 110°C. Lastly, the mortar and pestle were used to separate coated biomass powder. Heavy metal concentrations in a range from 1 to 100 mg/L were taken in order to evaluate the biosorption efficiency of the dry fungal cells. The heavy metal removal efficiency by *A. niger* biomass for different elements like zinc (Zn) II, mercury (Hg) II, fluorine (F) I, cobalt (Co) II, silver (Ag) I, and Cu were 100%, 83.2%, 83%, 71.4%, 48%, and 37%, respectively. The metal removal efficiency of *A. niger* biomass from the industrial wastes was evaluated in the presence of 5 g biomass, 95 mL nonsterile water contaminated with 250 mg/L Co

(II), 263 mg/L Pb (II), and 183 mg/L Hg (II), under stirring condition at 100 rpm, pH 5 and 28°C. The removal efficiencies of 96.4%, 69%, and 71% were noted at 7 days of incubation for the Co (II), Hg (II), and Pb (II), respectively.

Vacar et al. [30] reported the use of filamentous fungi as a potential bioremediators of Hg. The five fungal species were selected which were having the superior Hg resistance. The aqueous solution enriched with 100 mg/L Hg^{2+} was used for the biosorption by fungal species. Around 10^4 spores/mL were inoculated in the PDB medium (total volume 30 mL) and subjected to 7 days of incubation at 28°C and 120 rpm. The attained fungal biomass was vacuum filtered followed by washing and adding them into 100 mg/L aqueous solution of Hg^{2+}. Further, the mixture was incubated for 2 days under shaking conditions at 28°C and 120 rpm. The elimination efficiency of about 52% was noted for Hg^{2+} within 30 minutes in case of the fungal biomass whereas, *Didymella glomerata* P2.16 live biomass showed 93% elimination of Hg^{2+} after 2 hours of incubation period. All of the fungal isolates had acquired equilibrium at the end of 12 hours incubation period. The removal abilities of the live biomasses were $47\% \pm 8.0\%$ for *Sarocladium kiliense*, $62\% \pm 5.1\%$ for *F. oxysporum*, $56\% \pm 5.0\%$ for *Phoma costaricensis*, $61\% \pm 3.9\%$ for *Cladosporium* sp., and $97\% \pm 0.4\%$ for *D. glomerata*.

In another study, Li et al. [31] utilized immobilized fungal residues for investigating the adsorption ability toward Hg (II), Zn (II), and Cu (II). The immobilized fungal biomasses of *Pleurotus ostreatus*, *Auricularia polytricha*, *Flammulina velutipes*, and *Pleurotus eryngii* were prepared from the inedible stipe of mature sporophores by cutting and washing followed by drying at 60°C which were further ground, sieved, and stored at room temperature in a desiccator. A 25 mL metal ion solution was used in the batch experiments to evaluate the adsorption process. The highest adsorption of Hg^{2+}, Cu^{2+}, and Zn^{2+} was up to 5.18, 4.29, and 3.97 mg/g by *P. eryngii*, 6.32, 6.04, and 6.92 mg/g by *A. polytricha*, 5.18, 5.00, and 5.52 mg/g by *P. ostreatus* and 8.20, 8.13, and 7.23 mg/g by *F. velutipes*, respectively. The equilibrium adsorption capacities were 61.06%, 75.88%, 62.28%, and 81.74% for Cu^{2+}, 74.05%, 75.65%, 84.10%, and 80.17% for Zn^{2+}, 56.74%, 53.77%, 54.99%, and 74.68% for Hg^{2+} by *P. ostreatus*, *A. polytricha*, *F. velutipes*, and *P. eryngii*, respectively. First 60 minutes of adsorption accounted for the removal of a significant portion of the total metal content. The highest metal sorption affinity was noted for *F. velutipes*. The Hg^{2+}(0.02 mg/L), Zn^{2+} (0.01 mg/L), and Cu^{2+} (0.30 mg/L) adsorption from the water by *F. velutipes* was 69.35%, 66.67%, and 73.11%, respectively. The pseudo-first-order kinetic model was better suited for the metal ions sorption while the chemisorption process was denoted as the rate-limiting step. The functional groups such as carboxyl, phosphate, amide, and hydroxide present in the cell wall of adsorbents were considered as main contributors to the uptake of metal ions. The complexation reaction and the ion exchange among the adsorbent and functional groups was the major mechanism followed by the sorption process.

3.5   COPPER

Although copper is an essential element for maintaining the activity of several metabolic enzymes and associated reactions, a higher amount of the same can prove to be toxic [32]. Hence, attempts are made to remove hazardous levels of copper contaminant from wastewater using fungi. Rose and Devi [33] employed native filamentous fungi derived from the sludge and industrial wastewater for the bioaccumulation of heavy metals such as Ni (II) and Cu (II). Aliquots of about 1 mL from various dilutions were spread on Martin agar (MA), modified PDA, and modified Czapek yeast extract agar (CYEA) and media plates followed by 72 hours of incubation at 30°C. The visible heterogeneous colonies were purified and used for metal removal. A sterilized modified CYEA was inoculated with a loopful of fungal comprising 100 mg/L concentration of specific heavy metal. After the incubation for 72 hours or more at 30°C, the visible growth obtained was further subjected to further screening. The sterile modified CYEA plates containing 100 mg/L of heavy metals were inoculated with the fungus mycelium disk. The pH of the wastewater was around 3.08 which indicated its acidic nature. Cr, Pb, Cu, and Ni concentrations in the sludge sample were 89.23, 1.56, 19.34, and 60.86 mg/L, respectively while in the wastewater sample, it was 112.34, 2.94, 35.08, and 64.22 mg/L, respectively. MICs of Ni (II) and Cu (II) were found to be ranging from 2.5 to 5 mg/mL and 4 to 5.5 mg/mL, respectively. The tolerance of Cu (II) was observed in the following order *A. flavus* < *A. niger* < *A. awamori*. Similarly, for Ni (II), was *A. niger* < *A. flavus* ≈ *A. awamori*. The maximum growth (dry weight) and Cu (II) bioaccumulation was 10.41 g/L and 6.99 mg/g in *A. flavus* at pH 5, 9.46 g/L and 7.02 mg/g in *A. niger* at pH 4, and 12.30 g/L, and 7.13 mg/g *A. awamori*, respectively. In the case of Ni (II), the highest bioaccumulation quantity and growth was 9.50 g/L and 6.53 mg/g for *A. niger*, 12.31 g/L and 7.64 mg/g for *A. flavus*, and 10.19 g/L and 6.78 mg/g for *A. awamori*, respectively.

Pundir et al. [34] employed the Taguchi method for the elimination of Ni and Cu with the help of *Aspergillus* sp. The fungal biomass was cultivated in the growth media containing 5.0 g/L yeast extract, 0.5 g/L of $MgSO_4$, 0.5 g/L of NH_4NO_3, 0.5 g/L of K_2HPO_4, and 0.5 g/L of NaCl followed by its incubation under shaking conditions (180 rpm) at 30°C. The optimized parameters were 15% of inoculum concentration, 50 mg/L initial metal (Cu/Ni) concentration, pH 4, and 30°C. The maximum removal of Ni and Cu at the optimized conditions was 97.9% and 98.8%, respectively by the *Aspergillus* sp. biomass. The effect of parameters in the removal of Ni was in the order, temperature (0.10%) < pH (7.45%) < inoculum concentration (38.4%) < initial metal concentration (49%) while, in the case of Cu removal the order was temperature (0.83%) < pH (7.75%) < concentration of inoculum (30.40%) < initial concentration (45.62%).

In another study, Iskandar et al. [35] reported the biosorption and tolerance of filamentous fungi isolated from freshwater against Pb and Cu. The isolation of fungal species was carried out using sediment as the source. The preparation of the stock solution was carried out by the addition of around 10 g of sediment to the sterile distilled water. The solution was further kept under shaking conditions for 20 minutes at 200 rpm followed by the addition of 1 mL soil solution into each Petri dish and further pouring of 9 mL sterilized Rose Bengal agar (RBA) into each Petri dish which was subjected to 4 days of incubation at 28°C ± 2°C. The colonies that appeared were further sub-cultured onto PDA. The sediment used contained a high concentration of Pb (17.28 μg/g) and Cu (44.43 μg/g). The isolates were further classified into three genera; *Penicillium* sp., *Trichoderma* sp., and *Aspergillus* sp., and the isolates were identified as *P. janthinellum* (7 isolates), *T. asperellum* (9 isolates), *A. niger* (9 isolates), *A. fumigatus* (6 isolates), and *P. simplicissimum* (10 isolates). *P. simplicissimum* and *A. niger* were the only fungal isolates that tolerated and developed mycelium growth at 1,000 mg/L Cu (II) concentration whereas only *A. niger* showed growth and tolerance at 5,000 mg/L Pb (II) concentration. The highest uptake of Cu (II) (200 mg/L) was observed by *A. niger* (20.910 mg/g) followed by *P. simplicissimum* (16.179 mg/g), and *T. asperellum* (12.809 mg/g). Similarly, in the case of Pb (II) elimination, the maximum removal obtained by *A. niger* (54.047 mg/g) at 250 mg/L Pb concentration was followed by *P. simplicissimum* (38.975 mg/g) and *T. asperellum* (17.619 mg/g).

Vaseem et al. [36] used macro fungus *P. ostreatus* for the decontamination of pollution caused due to heavy metals from the coal washery effluent (CWE). Initially, CWE was diluted using distilled water (DW) in different combinations like 25% CWE + 75% DW and 50% CWE + 50% DW as well as only pure CWE. A mixture of diluted as well as pure CWE in equal amounts of malt dextrose agar (MDA) was prepared. The mycelium discs (1.5–2 cm) were added in the media comprising different CWE percentages which were further kept for 28 days for the growth of mycelia. The samples of both mycelia and media were further digested using diacid (2 HNO_3: 1$HClO_4$) at 130°C for analysis of the metal. The heavy metal remediation was maximum on the 20th day of the exposure period in 50% effluent and it was around 98% (Ni), 82% (Zn), 99.1% (Cr), 99.3% (Co), and 89.2% (Fe). In the case of raw effluent, the percentage reduction of 73% (Zn), 78.7% (Ni), 64.6% (Cr), 59.3% (Co), and 34.6% (Fe) was noted whereas, in the case of diluted effluent (25%) it was 55.1%, 97.8%, 87.06%, 84.5%, and 97% for Zn, Ni, Fe, Cr, and Co, respectively. The reduction (%) in Cu concentration on the 20th day was highest in 50% diluted effluent (99.9%) followed by that of 25% diluted effluent (99.7%) and raw effluent (87.5%). Thus, the efficiency of bioremediation was enhanced by the dilution of the effluent.

3.6 LEAD

Highly toxic lead (Pd) released during activities associated with mining, electroplating, and steel industries can result in severe health hazards like metal retardation and cardiovascular diseases. Efficient bioremediation processes are being searched for the removal of this extremely hazardous metal from the environment. Qiu et al. [37] investigated Pb^{2+} and Cd^{2+} immobilization using *A. niger*. The fungal strain was isolated and cultured in a PDA medium for sporulation after 5 days of incubation at 28°C. The spores formed were further recovered from the plates by scraping the surface. The mycelial fragments were eliminated via filtration and the spore count of around 8.9×10^7/mL was obtained. The experiments were carried out with five treatments like CK (control, no metals), Pb (0.893 mmol/L), Cd (0.893 mmol/L), 1Pb1Cd (Pb^{2+} and Cd^{2+} = 0.893 mmol/L) and 2Pb1Cd (Pb^{2+} = 1.786 mmol/L, Cd^{2+} = 0.893 mmol/L). The solid $Pb(NO_3)_2$ and $Cd(NO_3)_2$ powder were added to the PDB before the inoculation of the spore suspension into 100 mL PDB medium which was subjected to 5 days of incubation under shaking conditions at 28°C. The pH of the medium in the initial stage was around 6.5. A dramatic increase in the biomass from 0.2 to 0.8 g was noted during 3 days of incubation in CK and Pb treatments. A slight decrease to 0.7 g was noted after 5 days of incubation. The removal rates of Pb^{2+} were 98.74%, 99.15%, and 97.82% after 1, 3, and 5 days of incubation, respectively. The removal rates of Pb^{2+} were relatively stable in 1Pb1Cd treatment with removal efficiencies of 87.78%, 89.84%, and 88.73% on 1, 3, and 5 days of incubation, respectively. The surface morphology of mycelium as well as the adsorption of mineral particles was revealed from the SEM images.

Xu et al. [38] evaluated the elimination mechanism for Pb (II) employing *P. polonicum*. The medium was composed of 0.5 g $NaNO_3$, 5×10^{-3} g $MgSO_4 \cdot 7H_2O$, 2 g NaCl, 1 g beef extract, 0.1 g NH_4Cl, 3 g tryptone, 3 g glucose, 0.5 g yeast extract, and 0.5 g yeast extract. The reaction medium possessed acidic pH which was around 5. The fungus was grown in the medium at $30°C \pm 0.5°C$. Pb (II) removal efficiency of 90.3% was noted within 12 days by *P. polonicum* while the adsorption of 187.1 mg of Pb (II) was noted upon the treatment with the biomass. The experimental studies showed that the Pb (II) was immobilized in the form of lead oxalate exterior to the fungal cell, where it got bound with the nitro, phosphate, halide, amino, carboxyl, and hydroxyl groups associated with the cell wall. The metal got precipitated in the form of pyromorphite (cell wall) which was lastly reduced to Pb (0) inside the cell. Thus, the fungal strain *P. polonicum* showed Pb (II) tolerance at 4 mmol/L along with its successful elimination within 12 days.

Sharma et al. [39] studied the tolerance and removal efficiency of Pb using *A. terreus* (SEF-b) and *Talaromyces islandicus* (K1SF-Pb-15). The fungal strains were isolated from the industrial effluents, sludge, and sewage in the two strains were selected for further Pb adsorption experiments. On PDA slants, the selected fungal cultures were cultivated for 5 days and sterilized water (10 mL) was added to each slant after incubation. For preparing a spore suspension (1.4×10^6 spore/mL), the fungal growth was scrapped and appropriately diluted with sterilized DW. The fungal suspension was further inoculated into a flask (250 mL) containing 100 mg/L of Pb (II) with 100 mL PDB and the pH ranged from 2 to 7 followed by incubation under continuous shaking conditions at $30°C \pm 2°C$ for 5 days. The obtained fungal biomass was filtered by using Whatman filter paper No.1 after 5 days, and then it was dried in a hot oven for 48 hours at 80°C. The SEM images showed that both fungal hyphae were septate, branched, and cylindrical without any metal treatment after incubation for 120 hours. However, in the presence of 100 mg/L of Pb, there was complete disruption and breakdown of the fungal mycelium after incubation for 120 hours. The FTIR analysis of *T. islandicus* showed typical peaks at 2,924 cm^{-1} (methyl group stretching vibration), 3,500–3,200 cm^{-1} (N-H and -OH bond), and 1,373 cm^{-1} (CH- bending). Also, the spectra show a shift in peak from 1375.42 cm^{-1} (C=C) to 1438.35 cm^{-1} (C-O) and 3367.41 cm^{-1} (N-H) to 3397.40 cm^{-1} (O-H) after Pb biosorption. The spectra of *A. terreus* showed major peaks at 2,926 cm^{-1} (CH_2, aldehyde), 1,210 cm^{-1} (C–OH), and 3,422 cm^{-1} (O–H and N–H stretching). The peaks were shifted to 2,927, 1,300, and 3,430 cm^{-1} after Pb biosorption. The effect of pH was determined in the range 2–7. The optimal pH was 5 for the Pb uptake (4.56 mg/g) by *T. islandicus* with 88.1% removal efficiency. In case of *A terreus*, 90% removal efficiency with maximum uptake of 4.12 mg/g was noted at pH 5 as well. The optimization of the incubation time indicated maximum uptake of 3.73 mg/g with 90.06% removal efficiency for *T. islandicus* was obtained at 120 hours.

Likewise, maximum uptake (4.28 mg/g) and 90.74% removal efficiency were observed by *A. terreus* at 120 hours. The optimum inoculum size for both fungal strains was 4%. It was noted that the eradication efficiency decreased for both strains by increasing the metal concentration. The different carbon and nitrogen sources also affected the Pb removal percentage. The *T. islandicus* showed maximum percentage elimination and uptake of Pb when dextrose was used as a carbon source and ammonium chloride was utilized as a nitrogen source. On the other hand, a minimum percentage of removal and uptake was observed when starch was utilized as a carbon source. In case of *A. terreus*, the maximum percentage uptake and Pb removal were observed when lactose was used as a carbon source and peptone was used as the nitrogen source. Minimum percentage removal and Pb uptake were observed when starch was used as a carbon source and ammonium sulfate was used as a nitrogen source for *A. terreus*. Adsorption isotherm identified the Langmuir isotherm model as the best-fitted model compared to the Freundlich isotherm model, the maximum Pb biosorption capacity in *T. islandicus* was 8.62 mg/g ($R^2 = 0.98$), and *A. terreus* was 5.55 mg/g ($R^2 = 0.96$).

3.7 CHROMIUM

Chromium (Cr) is a toxic metal that is released in the effluents of the industries associated with electroplating, fertilizer, tannery, and textile. The two main oxidization forms are Cr (III) and Cr (VI) among which Cr (VI) is highly toxic and a potent carcinogen as well as a mutagen. Organs like kidney, liver, and lungs are affected by Cr (VI). Various fungi are employed for bioconversion and bioremoval of Cr. Pourkarim et al. [40] used the Artist's Bracket (AB) fungi for the removal of Cr (VI) from an aqueous solution. The AB was collected, washed, dried, and crushed followed by sieving (3.36 mm sieve). The recovered particles were washed with deionized water and dried for 24 hours at 105°C. The adsorption experiment was first started with the fresh stock solution of Cr (VI) prepared by dissolving a potassium dichromate (95%) in deionized water. Further, 100 mL of 50 mg/L Cr (VI) solution was poured into flasks, and the pH (range of 2, 4, 6, 8, and 10) was adjusted using HCl and NaOH solution. After adjusting the pH, the different doses of AB particles (0.05–0.5 g) were mixed with the prepared solution at different Cr (VI) concentrations (25, 50, 100, 200, and 300 mg/L) at 25°C. SEM images showed many pores with multi-layered walls on the surface of the biosorbent before Cr (VI) adsorption. The FTIR analysis revealed peaks of -OH (3,378.72 cm^{-1}), C=O (1,700.37 cm^{-1}), and C-N and C-O (1,293.25 cm^{-1}) which were present in AB particles. The bands observed at 1,537–1,540 cm^{-1} and 1,027–1,039 cm^{-1} were attributed to amide 2 and C–O stretching of alcohols and carboxylic acids, respectively. The optimum Cr (VI) adsorption was noted at pH 2. It is important to note that the Cr (VI) removal percentage was decreased with increasing the biosorbent concentration. The optimum biosorbent concentration for Cr (VI) removal was 50 mg/L. The maximum removal efficiency for Cr (VI) reached equilibrium within 30 minutes. The decrease in the Cr (VI) removal efficiency with the increase in the temperature and concentration of the Cr (VI) solution was noted. The kinetic study revealed that the pseudo-second-order model was a well-fitted compared to the pseudo-first-order model. The regression coefficients (R^2) were calculated from the Langmuir model (0.9998) (200 mg/g) and the Freundlich model (0.9992), indicating that both models were well-fitted models for the Cr (VI) sorption process.

In another study, Kumar and Dwivedi [19] studied Cr (VI) removal efficiency by the *A. flavus* CR500. The fungus was isolated from the electroplating wastewater. The fungal mycelia (CR500) which grew on the Cr (VI) amended PDA plate was further incubated for 6 days at 28°C ± 1°C. The metal removal experiment was carried out in a flask (250 mL) containing various concentrations of Cr (VI) ranging from 0 to 200 mg/L (0, 5, 20, 50, 100, and 200 mg/L) amended PDB at pH 6.5. A 1 mL of fungal spore suspension (3×10^4 mL) was further incubated in a continuous shaking condition at 28°C ± 1°C for 6 days at 80 rpm. The culture (5 mL) was collected from the flask after various incubation times of 24, 48, 72, 96, 120, and 144 hours followed by centrifugation at 10,000 rpm for 10 minutes and the supernatant was used to evaluate the Cr (VI) concentration. The Cr (VI) removal efficiency of isolate CR500 was studied at various temperatures (20°C, 28°C, 35°C, and 40°C) and pH (5, 6.5, 8, and 9.0) at 50 mg/L concentration of Cr (VI). SEM images showed the lack of globular protrusions in the mycelia after exposure to Cr (VI) while in the absence of Cr (VI) images showed roughly dense globular protrusions in the mycelia. Energy dispersive X-ray (EDX) analysis identified the presence of Cr (1.82 wt. %) at Cr (VI) concentration (100 mg/L). FTIR analysis revealed the extinction of peaks in Cr (VI) treated biomass at 1,744.6 and 1,374.3 cm^{-1} which were associated with the stretching vibration of C=O group and amide III bands that represent COO$^-$ group, respectively. The band at 2,134.8 cm^{-1} specific to the C≡C stretching also disappeared. The shift in the peaks at 3,407.5 cm^{-1} (-OH) and 3,375.6 cm^{-1} was attributed to the -NH stretching vibration of the protein which indicated the lone pair-metal interaction. The peaks at 1,554.9–1,549.0 cm^{-1}, and 1,648.9–1,650.5 cm^{-1}, indicated the protein-linked functional group to participate in the Cr (VI) complexation. It was speculated that the Cr was removed via precipitation/adsorption and accumulation on the surface of mycelia. The CR500 showed efficient Cr (VI) removal ability between pH 5 and 9, while the highest of 99.4% Cr (VI) reduction was observed at pH 6.5. The CR500 was capable of decreasing Cr (VI) (50 mg/L) at temperatures ranging from 28°C to 40°C. The CR500 was able to tolerate 800 mg/L of Cr (VI). The Cr (VI) reduction capacity of CR500 was 89.1% which was achieved in 24 hours at 4.9 ± 0.12 mg of Cr per gram of dried biomass accumulation. The highest Cr (VI) removal of 79.4% was observed at 5 mg/L in 144 incubation time. The phytotoxicity test for fungus-treated and untreated Cr (VI) containing solution (100 mg/L) was checked using *Vigna radiata* seeds. The CR500 effectively detoxified the Cr (VI) via removal and reduction mechanisms. The chromate reductase (ChrR) enzyme was responsible for the reduction of Cr (VI) to Cr (III) by CR500, where GSH (glutathione), glucose, NADH, and acetate played the role of electron donors.

Da Rocha Ferreira et al. [41] used the *Pleurotus ostreatus* fungi for the elimination of Cr (VI) from wastewater. The removal efficiency was checked with two different media: solid medium and liquid medium with inactive and active fungal biomass. The experiment in the solid medium was initiated with the *P. ostreatus* strain grown on a PDA medium for 7 days at 28°C ± 2°C where the pH was adjusted to 5.6. After incubation, the fungal colonies grew on various Cr (VI) concentrations (150, 125, 100, 75, 50, 25, and 10 mg/L) containing agar plates. The active biomass was prepared in PDA at $28°C \pm 2°C$ for 14 days and was further divided into two parts. One part used inactive fungus for the biosorption study, and the other part used active fungus for the biosorption study. The inactive biomass was obtained by filtering and washing the biomass with DW followed by drying for 24 hours at $80°C \pm 2°C$ in an oven. Further, the obtained biomass was ground and sieved (36-mesh). Biosorption by inactive and active fungal biomass was examined in the liquid medium in which flasks contained 100 mL medium (22 g/L of glucose, 13.2 g/L of $NaNO_3$, 3.3 g/L of KH_2PO_4, 3.3 g/L of KCl, 1.1 g/L of $MgSO_4 \cdot 7H_2O$, 0.0022 g/L of $FeSO_4$ and 0.0022 g/L of $ZnSO_4$. Each flask had different Cr (VI) concentrations equivalent to 150, 125, 100, 75, 50, 25, and 10 mg/L. In the active biomass-mediated treatment, 1 mL aliquot of fungal suspension was added in each conical flask while treatment with the inactive biomass was initiated with addition of 0.2 g of fungal biomass in each conical flask. The inactive and active biomass-mediated treatment was performed under shaking condition (150 rpm) at $28°C \pm 2°C$. A 2 mL of aliquot was collected at every 24 hours for 15 days for all concentrations in case of active biomass-treated samples. In case of the inactive fungal biomass-mediated treatment, 2 mL of samples were recovered every 2 minutes for 22 minutes. The absorbance was observed using a UV–visible spectrophotometer at 540 nm. In case of the Cr (VI) adsorption by active biomass, the 100% of biosorption was reached in 360 hours at a Cr (VI) concentration of 25 mg/L, which was equal to the elimination of 169.84 mg/g of total Cr. In case of the inactive biomass, the 100% of biosorption was reached in 22 minutes at a Cr (VI) concentration of 50 mg/L, which was equivalent to the removal of 368.21 mg/g of total Cr.

Espinoza-Sanchez et al. [42] used dead fungal biomass of *Rhizopus* spp. for the removal of Cr (VI) in an aqueous solution. The fungal isolates were grown in a 200 mL flask containing media composed of 40 g/L saccharose, 75 g/L $MgSO_4 \cdot 7H_2O$, 4 g/L K_2HPO_4, 1 g/L NH_4Cl, 3 g/L citric acid, 1.5 g/L yeast extract, 0.5 g/L cysteine, 0.1 g/L $CaCl_2$, and 1 g/L NaCl which were further incubated in a shaker for 7 days at 30°C. The resulting biomass was filtrated using vacuum filter units made up of cellulose acetate membrane with 0.22 μm pores. Further, the NaCl (0.8 M) solution was used for the washing of half of the obtained fungal biomass followed by washing with deionized water. The biomass was inactivated thereafter by autoclaving at 121°C at 1.0207 atm followed by drying for 24 hours at 60°C in the oven. The particles of size ~10 μm were generated using a mortar and pestle. The batch adsorption experiment was screened with 2^3 experimental designs: the experiment was carried out by taking 10 mL Cr (VI) concentration (50 mg/L) at pH 2–4, with the 50 mg/L dosage of fungal biomass for 120 minutes in an orbital shaker at 50 and 150 rpm and with temperatures of 20°C–30°C, followed by centrifugation. The kinetic experiments were carried out with 10 mL of 50 mg/L Cr (VI), 120 rpm, $25 \pm 2°C$, pH 2.0, and 0.05 g of biomass dosage with various contact times ranging from 0.25 to 8.0 hours. The SEM images showed the plaster-like structure after Cr (VI) adsorption. The EDX analysis exhibited Na and Cl before adsorption which decreased after adsorption. The EDX analysis confirmed that after adsorption, the Cr was localized in the biomass as evident from Figure 3.2. The FTIR analysis revealed the participation of

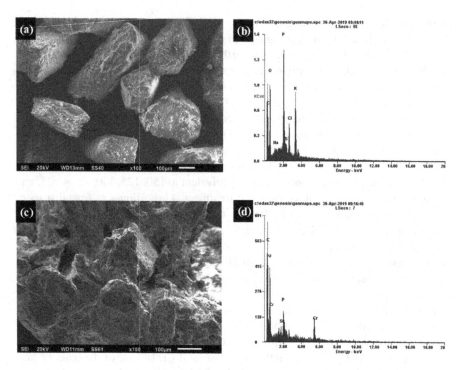

FIGURE 3.2 SEM images before (a) and after (c) adsorption, (b) and (d) correspond to the respective EDX spectra. Reprinted with permission from Espinoza-Sanchez, M.A., Arevalo-Nino, K., Quintero-Zapata, I., Castro-Gonzalez, I., Almaguer-Cantu, V., 2019. Cr (VI) adsorption from aqueous solution by fungal bioremediation based using *Rhizopus* sp. *J. Environ. Manage.* 251, 109595. Copyright © 2019 Elsevier Ltd.

amino groups (3,275 cm^{-1}) in the adsorption of Cr (VI). The bands were observed at 2,922 and 2,853 cm^{-1} attributed to the –CH group, at 1,713 cm^{-1} (C=O of the carboxyl groups), 1,622 cm^{-1} (amide bond corresponding to a C=O stretch), and the band at 1,065 and 1,031 cm^{-1} (-PO$_4$$^{3-}$). The maximum eradication of Cr (VI) was 90% and 99% by dead biomass of *Rhizopus* sp. at pH 4.0 and 2.0, respectively, which indicated the highest adsorption was achieved at an acidic pH due to the Cr (VI) three ionic states (HCrO$_4$$^{-}$, Cr$_2O_7$$^{2-}$, and CrO$_4$$^{2-}$). This increased the electrostatic attraction between the sorbate and the sorbent. The maximum Cr (VI) adsorption equilibrium was reached in 240 minutes for both treated and untreated biomass. The kinetic study revealed the pseudo-second-order model ($R^2 > 0.99$) was well-fitted and the Elovich model ($R^2 = 0.99$) in a non-linear form, signifying chemisorption was the controlling step of adsorption. The thermodynamic study revealed the adsorption process was spontaneous (negative $\Delta G°$) and endothermic ($\Delta G°$). The Langmuir isotherm model was best-fitted for the maximum Cr (VI) removal up to 9.95 mg/g by *Rhizopus* sp. + NaCl (Table 3.1).

TABLE 3.1
Metal Removal Efficiency of Fungi

Fungal Strain	Toxic Metal	Metal Ion Concentration	Removal Efficiency	References
Fusarium solani	Cd	50 mg/L	87.5% ± 4.2%	Kumar et al. [21]
Penicilliumchrysogenum, Aspergillus ustus	Cd	1 ppm	91% (*P. chrysogenum*) and 84% (*A. ustus*)	Alothman et al. [22]
Trichoderma lixii, Trichoderma harzianum, Trichoderma simmonsii, Trichoderma longibrachiatum, Trichodermacitrinoviride, Piriformosporaindica	Cd	5 mg/L (*T. lixii*), 10 and 500 mg/L (*T. simmosii*), 5, 50, and 200 mg/L (*P. indica*)	95.3% (*T. lixii*), 91.7% and 31.2% (*T. simmonsii*), 58.8%, 2.98%, and 0.18% (*P. indica*)	Yaghoubian et al. [23]
Waste fungal biomass (WFB)	Cd	-	-	Rozman et al. [24]
Aspergillus flavus CR500	As	10 mg/L	99%	Kumar and Dwivedi[19]
Aspergillus flavus III, IV, and V, *Aspergillus fumigatus*I-II, *Paecilomyces* sp., *Cladosporium* sp., *Mucor* sp-1 and 2	As	-	99.3% (*Aspergillus flavus*IV), 92.3% (*Aspergillus flavus*III), 90.3% (*Aspergillus flavus*V), and 89% (*Paecilomyces* sp.)	Cardenas-Gonzalez et al. [26]
Saccharomyces cerevisiae	As	95 µg/L	78% (As (III)), 19.7% (As (V))	Hadiani et al. [27]
Trichoderma sp.	As	-	59%	Govarthanan et al. [28]
Aspergillus niger	Hg	183 mg/L	96.4%	Acosta-Rodriguez et al. [29]
Didymellaglomerata, Fusarium oxysporum, Cladosporium sp., *Phomacostaricensis, Sarocladiumkiliense*	Hg	100 mg/L	97% ± 0.4% (*D. glomerata*) > 62% ± 5.1% (*Fusarium oxysporum*) = 61% ± 3.9% (*Cladosporium* sp.) > 56% ± 5.0% (*Phomacostaricensis*) > 47% ± 8.0% (*Sarocladiumkiliense*).	Vacar et al. [30]
Flammulinavelutipes, Auriculariapolytricha, Pleurotuseryngii, Pleurotusostreatus	Hg	-	54.99% (*F. velutipes*), 53.77% (*A.*polytricha), 74.68% (*P. eryngii*), and 56.74% (*P. ostreatus*)	Li et al. [31]

(Continued)

TABLE 3.1 *(Continued)*
Metal Removal Efficiency of Fungi

Fungal Strain	Toxic Metal	Metal Ion Concentration	Removal Efficiency	References
Aspergillus awamori, Aspergillus niger,Aspergillus flavus	Cu	-	12.30 g/L and 7.13 mg/g (*A. awamori*),9.46 g/L and 7.02 mg/g (*A. niger*) and 10.41 g/L and 6.99 mg/g (*A. flavus*).	Rose and Devi [33]
Aspergillus sp.	Cu	50 mg/L	98.8%	Pundir et al. [34]
Aspergillus niger, Trichoderma asperellum,P enicilliumsimplicissimum	Cu	250 mg/L	54.047 mg/g (*A. niger*), 38.975 mg/g (*P. simplicissimum*) and 17.619 mg/g (*T. asperellum*)	Iskandar et al. [35]
Pleurotusostreatus	Cu		99.9%	Vaseem et al. [36]
Aspergillus niger	Pb	-	98.7%	Qiu et al. [37]
Penicilliumpolonicum	Pb	-	90.3%	Xu et al. [38]
Talaromycesislandicus (K1SF-Pb-15), *Aspergillus terreus* (SEF-b)	Pb	-	90%	Sharma et al. [39]
Artist's Bracket	Cr	-	200 mg/g	Pourkarim et al. [40]
Aspergillus flavus CR500	Cr	50 mg/L	89.1%	Kumar and Dwivedi[19]
Pleurotusostreatus	Cr	25 mg/L (active biomass), 50 mg/L (inactive biomass)	169.84 mg/g (active biomass), 368.21 mg/g (inactive biomass)	Da Rocha Ferreira et al. [41]
Rhizopus spp.	Cr	50 mg/L	99%	Espinoza-Sanchez et al. [42]

3.8   SUMMARY AND CONCLUSION

Fungi-mediated removal of hazardous heavy metals is an efficient green approach that can immensely benefit the industrial effluent treatment processes [43,44]. It would not only save time but also bring down the cost of the overall process as it doesn't require the addition of any chemicals for precipitation unlike the chemical treatment processes. Moreover, the used fungal biomass from the bakery and vineries can be reused for metal removal as even the dead mycelia are capable of absorbing heavy metals. However, various parameters such as type of fungi, inoculum density, contact time, concentration of metal, nutrient supplements, and agitation should be carefully optimized to get the maximum benefit of the process [45,46]. Certain fungal enzymes such as chromate reductase (ChrR) help in the bioconversion process. Such enzymes can be immobilized on nanoparticles for recycle and reuse in the process [47–49]. The underlying molecular mechanism behind the process should be thoroughly studied using an integrated approach involving genomics, proteomics, and metabolomics. The involved genes can be introduced to other microbes to develop genetically modified heavy metal-removing organisms.

In view of the background, it can be concluded that fungi-associated heavy metal remediation is a significant step toward sustainable wastewater treatment. Rational studies toward optimization and scale-up would certainly help in implementing the process from laboratory to community scale.

REFERENCES

1. Fu, F., Wang, Q., 2011. Removal of heavy metal ions from wastewaters: A review. *J. Environ. Manage.* 92(3), 407–418.
2. Demir, A., Arisoy, M., 2007. Biological and chemical removal of Cr (VI) from waste water: Cost and benefit analysis. *J. Hazard. Mater.* 147(1–2), 275–280.
3. Mondal, N. K., Samanta, A., Dutta, S., Chattoraj, S., 2017. Optimization of Cr(VI) biosorption onto *Aspergillus niger* using 3-level box- behnken design: Equilibrium, kinetic, thermodynamic and regeneration studies. *J. Genet. Eng. Biotechnol.* 15(1), 151–160.
4. Ali, Z., Malik, R. N., Shinwari, Z. K., Qadir, A., 2015. Enrichment, risk assessment, and statistical apportionment of heavy metals in tannery-affected areas. *Int. J. Environ. Sci. Technol.* 12, 537–550.
5. Ayangbenro, A. S., Babalola, O. O., 2017. A new strategy for heavy metal polluted environments: A review of microbial biosorbents. *Int. J. Environ. Res. Public. Health.* 14(1), 94.
6. Khan, I., Aftab, M., Shakir, S., Ali, M., Qayyum, S., Rehman, M. U., Touseef, I., 2019. Mycoremediation of heavy metal (Cd and Cr)-polluted soil through indigenous metallotolerant fungal isolates. *Environ. Monit. Assess.* 191, 585.
7. Jin, Y., Luan, Y., Ning, Y., Wang, L., 2018. Effects and mechanisms of microbial remediation of heavy metals in soil: A critical review. *Appl. Sci.* 8(8), 1336.
8. Haq, I. and Kalmdhad, A.S. 2023. Enhanced biodegradation of toxic pollutants from paper industry wastewater using Pseudomonas sp. immobilized in composite biocarriers and its toxicity evaluation. *Bioresour. Technol. Rep*, 24, 101674.
9. I. Haq, A.S. Kalamdhad, A. Pandey, 2022. Genotoxicity evaluation of paper industry wastewater prior and post-treatment with laccase producing Pseudomonas putida MTCC 7525. *J. Clean. Prod.*, 342, 130981.
10. I. Haq, A.S. Kalamdhad, 2021. Phytotoxicity and cyto-genotoxicity evaluation of organic and inorganic pollutants containing petroleum refinery wastewater using plant bioassay. *Environ. Technol. Innov.*, 23, 101651.
11. I. Haq, A. Raj, 2018. Markandeya Biodegradation of Azure-B dye by Serratia liquefaciens and its validation by phytotoxicity, genotoxicity and cytotoxicity studies. *Chemosphere*, 196, 58–68.
12. I. Haq, S. Kumar, A. Raj, M. Lohani, G.N.V. Satyanarayana, 2017. Genotoxicity assessment of pulp and paper mill effluent before and after bacterial degradation using Allium cepa test. *Chemosphere*, 169, 642–650.
13. I. Haq, S. Kumar, V. Kumari, S.K. Singh, A. Raj, 2016. Evaluation of bioremediation potentiality of ligninolytic *Serratia liquefaciens* for detoxification of pulp and paper mill effluent. *J. Hazard Mater.*, 305, 190–199.

14. I. Haq, V. Kumari, S. Kumar, A. Raj, M. Lohani, R.N. Bhargava, 2016. Evaluation of the phytotoxic and genotoxic potential of pulp and paper mill effluent Using Vigna radiata and Allium cepa. *Adv. Biol.*, 2016, 1–10, Article ID 8065736.

15. Siddiquee, S., Rovina, K., Azad, S. A., Naher, L., Suryani, S., Chaikaew, P., 2015. Heavy metal contaminants removal from wastewater using the potential filamentous fungi biomass: A review. *J. Microb. Biochem. Technol.* 7(6), 384–395.

16. Sharma, S., Rana, S., Thakkar, A., Baldi, A., Murthy, R. S. R., Sharma, R. K., 2016. Physical, chemical and phytoremediation technique for removal of heavy metals. *J. Heavy Met. Toxic. Dis.* 1(2), 10.

17. Sakthivel, M., Ayyasamy, P. M., Hemalatha, N., 2016. Biotransformation of chromium (Cr-VI to Cr III) from tannery effluent using bacteria and fungi. *Int. J. Adv. Res. Biol. Sci.* 3(11), 68–79.

18. Xiong, J. Q., Kim, S. J., Kurade, M. B., Govindwar, S., Abou-Shanab, R. A., Kim, J. R., Jeon, B. H., 2019. Combined effects of sulfamethazine and sulfamethoxazole on a freshwater microalga, *Scenedesmus obliquus*: Toxicity, biodegradation, and metabolic fate. *J. Hazard. Mater.* 370, 138–146.

19. Kumar, V., Dwivedi, S. K., 2019. Hexavalent chromium reduction ability and bioremediation potential of *Aspergillus flavus* CR500 isolated from electroplating wastewater. *Chemosphere* 237, 124567.

20. Gola, D., Dey, P., Bhattacharya, A., Mishra, A., Malik, A., Namburath, M., Ahammad, S. Z., 2016. Multiple heavy metal removal using an entomopathogenic fungi *Beauveria bassiana*. *Bioresour. Technol.* 218, 388–396.

21. Kumar, V., Singh, S., Singh, G., Dwivedi, S. K., 2020. Exploring the cadmium tolerance and removal capability of a filamentous fungus *Fusarium solani*. *Geomicrobiol. J.* 36, 782–791.

22. Alothman, Z. A., Bahkali, A. H., Khiyami, M. A., Alfadul, S. M., Wabaidur, S. M., Alam, M., Alfarhan, B. Z., 2020. Low cost biosorbents from fungi for heavy metals removal from wastewater. *Sep. Sci. Technol.* 55, 1766–1775.

23. Yaghoubian, Y., Siadat, S. A., Moradi Telavat, M. R., Pirdashti, H., Yaghoubian, I., 2019. Bio-removal of cadmium from aqueous solutions by filamentous fungi: *Trichoderma* spp. and *Piriformospora indica*. *Environ. Sci. Pollut. Res.* 26, 7863–7872.

24. Rozman, U., Kalcikova, G., Marolt, G., Skalar, T., Gotvajn, A. Z., 2020. Potential of waste fungal biomass for lead and cadmium removal: Characterization, biosorption kinetic and isotherm studies. *Environ. Technol. Innov.* 18, 100742.

25. Kumar, V., Dwivedi, S. K., 2020. Multimetal tolerant fungus *Aspergillus flavus* CR500 with remarkable stress response, simultaneous multiple metal/loid removal ability and bioremediation potential of wastewater. *Environ. Technol. Innov.* 20, 101075.

26. Cardenas-Gonzalez, J. F., Acosta-Rodriguez, I., Teran-Figueroa, Y., Rodriguez-Perez, A. S., 2017. Bioremoval of arsenic (V) from aqueous solutions by chemically modified fungal biomass. *3 Biotech*. 7, 226.

27. Hadiani, M. R., Khosravi-Darani, K., Rahimifard, N., 2019. Optimization of As (III) and As (V) removal by *Saccharomyces cerevisiae* biomass for biosorption of critical levels in the food and water resources. *J. Environ. Chem. Eng.* 7(2), 102949.

28. Govarthanan, M., Mythili, R., Kamala-Kannan, S., Selvankumar, T., Srinivasan, P., Kim, H., 2019. In-vitro bio-mineralization of arsenic and lead from aqueous solution and soil by wood rot fungus, *Trichoderma* sp. *Ecotoxicol. Environ. Saf.* 174, 699–705.

29. Acosta-Rodriguez, I., Cardenas-Gonzalez, J. F., Rodriguez Perez, A. S., Oviedo, J. T., Martinez-Juarez, V. M., 2018. Bioremoval of different heavy metals by the resistant fungal strain *Aspergillus niger*. *Bioinorg. Chem. Appl.* 2018, 1–7.

30. Vacar, C. L., Covaci, E., Chakraborty, S., Li, B., Weindorf, D. C., Frenţiu, T., Podar, D., 2021. Heavy metal-resistant filamentous fungi as potential mercury bioremediators. *J. Fungi*. 7(5), 386.

31. Li, X., Zhang, D., Sheng, F., Qing, H., 2018. Adsorption characteristics of copper (II), zinc (II) and mercury (II) by four kinds of immobilized fungi residues. *Ecotoxicol. Environ. Saf.* 147, 357–366.
32. Ghosh, S., 2018. Copper and palladium nanostructures: A bacteriogenic approach. *Appl. Microbiol. Biotechnol.* 102, 7693–7701.
33. Rose, P. K., Devi, R., 2018. Heavy metal tolerance and adaptability assessment of indigenous filamentous fungi isolated from industrial wastewater and sludge samples. *Beni-Suef Univ. J. Basic. Appl. Sci.* 7(4), 688–694.
34. Pundir, R., Chary, G. H. V. C., Dastidar, M. G., 2018. Application of Taguchi method for optimizing the process parameters for the removal of copper and nickel by growing *Aspergillus* sp. *Water Resour. Ind.* 20, 83–92.
35. Iskandar, N. L., Zainudin, N. A. I. M., Tan, S. G., 2011. Tolerance and biosorption of copper (Cu) and lead (Pb) by filamentous fungi isolated from a freshwater ecosystem. *J. Environ. Sci.* 23(5), 824–830.
36. Vaseem, H., Singh, V. K., Singh, M. P., 2017. Heavy metal pollution due to coal washery effluent and its decontamination using a macrofungus, *Pleurotus ostreatus*. *Ecotoxicol. Environ. Saf.* 145, 42–49.
37. Qiu, J., Song, X., Li, S., Zhu, B., Chen, Y., Zhang, L., Li, Z., 2021. Experimental and modeling studies of competitive Pb (II) and Cd (II) bioaccumulation by *Aspergillus niger*. *Appl. Microbiol. Biotechnol.* 105, 6477–6488.
38. Xu, X., Hao, R., Xu, H., Lu, A., 2020. Removal mechanism of Pb (II) by *Penicillium polonicum*: Immobilization, adsorption, and bioaccumulation. *Sci. Rep.* 10, 9079.
39. Sharma, R., Talukdar, D., Bhardwaj, S., Jaglan, S., Kumar, R., Kumar, R., Umar, A., 2020. Bioremediation potential of novel fungal species isolated from wastewater for the removal of lead from liquid medium. *Environ. Technol. Innov.* 18, 100757.
40. Pourkarim, S., Ostovar, F., Mahdavianpour, M., Moslemzadeh, M., 2017. Adsorption of chromium (VI) from aqueous solution by Artist's Bracket fungi. *Sep. Sci. Technol.* 52(10), 1733–1741.
41. Da Rocha Ferreira, G. L., Vendruscolo, F., Antoniosi Filho, N. R., 2019. Biosorption of hexavalent chromium by *Pleurotus ostreatus*. *Heliyon.* 5(3), e01450.
42. Espinoza-Sanchez, M. A., Arevalo-Nino, K., Quintero-Zapata, I., Castro-Gonzalez, I., Almaguer-Cantu, V., 2019. Cr (VI) adsorption from aqueous solution by fungal bioremediation based using *Rhizopus* sp. *J. Environ. Manage.* 251, 109595.
43. Shende, S., Joshi, K.A., Kulkarni, A.S., Charolkar, C., Shinde, V.S., Parihar, V.S., Kitture, R., Banerjee, K., Kamble, N., Bellare, J., Ghosh, S., 2018. *Platanus orientalis* leaf mediated rapid synthesis of catalytic gold and silver nanoparticles. *J. Nanomed. Nanotechnol.* 9(2), 494.
44. Shende, S., Joshi, K.A., Kulkarni, A.S., Shinde, V.S., Parihar, V.S., Kitture, R., Banerjee, K., Kamble, N., Bellare, J., Ghosh, S., 2017. *Litchi chinensis* peel: A novel source for synthesis of gold and silver nanocatalysts. *Glob. J. Nanomed.* 3(1), 555603.
45. Bloch, K., Mohammed, S.M., Karmakar, S., Shukla, S., Asok, A., Banerjee, K., Patil-Sawant, R., Mohd Kaus, N.H., Thongmee, S., Ghosh, S., 2022. Catalytic dye degradation by novel phytofabricated silver/zinc oxide composites. *Front. Chem.* 10, 1013077.
46. Gami, B., Bloch, K., Mohammed, S.M., Karmakar, S., Shukla, S., Asok, A., Thongmee, S., Ghosh, S., 2022. *Leucophyllum frutescens* mediated synthesis of silver and gold nanoparticles for catalytic dye degradation. *Front. Chem.* 10, 932416.
47. Robkhob, P., Ghosh, S., Bellare, J., Jamdade, D., Tang, I.M., Thongmee, S., 2020. Effect of silver doping on antidiabetic and antioxidant potential of ZnO nanorods. *J. Trace Elem. Med. Biol.* 58, 126448.

48. Karmakar, S., Ghosh, S., Kumbhakar, P., 2020. Enhanced sunlight driven photocatalytic and antibacterial activity of flower-like ZnO@MoS2 nanocomposite. *J. Nanopart. Res.* 22, 11.

49. Adersh, A., Kulkarni, A.R., Ghosh, S., More, P., Chopade, B.A., Gandhi, M.N., 2015. Surface defect rich ZnO quantum dots as antioxidant inhibiting α-amylase and α-glucosidase: A potential anti-diabetic nanomedicine. *J. Mater. Chem. B* 3(22), 4597–4606.

4 Biological Sulfate Conversion
Role of Electron Donors

Aparna Yadu, Anand Kumar J,
and Biju Prava Sahariah

4.1 INTRODUCTION

In the natural environment, elemental sulfur occurs as native sulfur whereas the most stable form of sulfur is sulfate, also found as sulfide, sulfite, thiosulfate, and various polythionates (e.g., tetrathionate) minerals (Liamleam & Annachhatre, 2007). Sulfate naturally exists in rocks, soil, and easily dissolves in water. Sulfide minerals available as sulfur compounds are pyrite (iron sulfide), sphalerite (zinc sulfide), galena (lead sulfide), cinnabar (mercury sulfide), and stibnite (antimony sulfide). In addition to natural sources of sulfate, industrial activities are also responsible for the formation of sulfate, for example, tannery, pulp and paper, textile, galvanization process, flue gas desulfurization, seawater intrusion, molasses fermentation, pharmaceutical production, food production, potato starch factories, tanneries, and petrochemical process (Klok et al., 2012). According to USEPA standards, the maximum permissible limit of sulfate in drinking water is 250 mg/L (Burns et al., 2012). The sulfur cycle is disturbed in the presence of high sulfate concentration which affects biodegradation pathways as well as decomposition rates and stimulates nutrient release from sediments as internal eutrophication and organic soil biodegradation (Geurts et al., 2009). Also, sulfide binds metal ions and sequesters them in the soil as metal sulfate. Acid mine drainage (AMD) or acid rock drainage is a rich source of sulfur compound released from abandoned mines, tailing piles or water stored in mines during rainy season (Bernardes de Souja & Mansur, 2011). Due to the increasing demand for various kinds of elements in the industrial sector, the abundance of minerals in mines is getting reduced nowadays and so many mines are now turned to be empty or useless enhancing AMD. Most of the metals found in environment in their sulfide form are FeS_2, CuS, and ZnS. For commercial exploitation these sulfide minerals are oxidized in the presence of air and water, and thus acidic sulfate-containing water is formed. The formation of AMD is based on the process of exposure of sulfide minerals mainly, iron sulfide to the environment, between sulfide minerals, water, and oxygen so that it produces sulfates, sulfuric acid, dissolved heavy metals and metal precipitates. The microorganism responsible for this reaction is known as

DOI: 10.1201/9781003368120-4

sulfur-oxidizing bacteria such as *Acidothiobacillus ferroxidans* (Alena and Maria, 2005). The following reaction represents the formation of AMD:

$$4FeS_2 + 14H_2O + 15O_2 \rightarrow 4Fe(OH)_3 + 8SO_4^{2-} + 16H^+$$

Several physicochemical and biological treatment methods are available for the removal of sulfate from metal and sulfate-containing wastewater. In physicochemical treatment, the most widely used for neutralization of acidic metal and sulfate-rich wastewater is the addition of neutralizing agent (Coulton et al., 2003). The very often neutralizing agents used are calcium oxide, calcium carbonate, sodium hydroxide, and lime. This neutralizing agent increases the pH as well causes the precipitation of metals and it accelerates the rate of chemical oxidation. However, on large scale, it is not cost effective. The use of anoxic limestone drains is another useful technology called armoring for in situ remediation of AMD where alkali is added to precipitate the metals on limestone (Johnson & Hallberg, 2005). However, armoring of metal precipitates retards the reactivity of limestones. In the case of in situ remediation on a large scale, sulfide passivation or micro-encapsulation using chemical compounds (Furfuryl alcohol resins) is one of the favorable techniques for sulfate remediation. Moreover, ionic exchange, nano-filtration and reverse osmosis can also be used for sulfate and metal removal (*Geotechnolgy I*, 2014 Springer). Sulfate reduction by physico-chemical treatment is an effective technology but production of the bulky sludge and high operating cost is the main disadvantages. Therefore, biological sulfate reduction with the help of SRB is considered to be a promising alternative for the treatment of metal and sulfate-bearing wastewater (Kieu et al., 2011). This process use microorganisms that can harness required amount of energy from redox reactions and sulfate works as a source of electron acceptor. The neutralization and precipitation of heavy metals viz, iron, copper, lead, mercury, etc. from mine water consume protons in the process of production of alkalinity and hydrogen sulfide.

$$SO_4^{2-} + 8e^- + 8H^+ \rightarrow H_2S + 2H_2O + 2OH^-$$

Moreover, the precipitation of metal sulfides leads to the release of protons.

$$M^{2+} + HS^- \leftrightarrow MS(s) + H^+$$

The diverse sulfate-reducing organism exists in industrial, mine, water, soil, and sediments, marine and freshwater etc. with capability to produce sulfide and bicarbonate in presence of suitable electron donor (Singh et al., 2011). Biological sulfate reduction under anaerobic condition is a widely used method for removal of sulfate and metal from wastewater. Research has focused on removal of sulfate and metal bearing wastewater by using anaerobic upflow sludge blanket reactor, anaerobic packed bed reactor or combination of sequential anaerobic-aerobic bioreactor and three stage sulfidogenic fluidized bed reactor (Silva et al., 2002; Jong and Parry, 2003; Ucar et al., 2011). Some sulfate-rich wastewaters possess inadequate amount of electron donors to achieve efficient sulfate removal process and demand external addition of electron donors. The objective of this article is to review the diversity of electron donors available for efficient biological sulfate reduction on the basis of their suitability and application.

4.2 MICROORGANISM INVOLVED IN SULFATE REDUCTION WITH THEIR MECHANISM PROCESS

Sulfate-reducing bacteria is a group of heterogeneous microorganism having ability to treat sulfate and metal in presence of suitable electron donor using sulfate as a terminal electron acceptor. Sulfate-reducing microorganism have diversity in nature such as *Deltaproteobacteria, Clostridia, Thermodesulfobacteria, Thermodesulfobiaceae, Nitrospirae, Euryarchaeota and Crenarchaeota*, etc. (Muyzer and Stams, 2008; Castro et al., 2000). The essential nutrients like carbon, nitrogen, oxygen and hydrogen have great impact on the growth of SRB in mixed culture and their biodegradation capability. Numerous literatures reported the significant effect of different electron donor on performance of SRB (Zhao et al., 2010; Wang et al., 2011; Hao et al., 2013; Cao et al., 2012). SRBs also be distinguishable in terms of their ability to oxidize organic substances either completely to CO_2 (*Desulfobacter, Desulfonema, Desulfobacterium*) or incompletely mainly to acetate (*Desulfomicrobium, Desulfovibrio Desulfotomaculum*) (Gibson, 1990; Muyzer and Stams, 2008). Table 4.1 presents the genus of sulfate-reducing microorganism involved in the treatment of sulfate and metal containing wastewater. In spite of that, the end product in the dissimilatory reduction of sulfate is sulfur and hydrogen sulfide.

TABLE 4.1
SRB Involved in the Treatment of Sulfate and Metal Containing Wastewater

Species	Carbon Source	Bioreactor	HRT	Temp.	Reference
Desulfovibrio, acetobacterium	H_2 & CO_2	Gas lift reactor	-	30°C	Houten et al. (2006)
Desulfosarcina variabilis	Ethanol, isopropanol & butanol	Lab-scale reactors	-	35°C	Dar et al. (2007)
Deltaproteobacteria, Desulfovibrio spp.,	Ethanol	Anaerobic sequencing batch biofilm reactor	48 hours	31°C ± 2°C	Sarti et al. (2010)
Desulfonema, Desulfobulbus, spp., and Desulfobacteriacea	Acetate	Continuous bioreactor	2.5 days	35°C	Icgen et al. (2007)
Desulfomicrobium spp., Desulfomicrobium	Ethanol and acetate	Gas lift reactor	4–7 days	30–35°C	Houten et al. (2006)
Desulfovibrio sp.	methanol	Fixed-bed reactor	20 hours	30°C	Glombitza (2001)
Thermodesulfobium spp., Desulfosporosinus spp., Desulfitobacterium spp., Desulfotomaculum spp.	Glycerol, methanol, hydrogen	In situ	-	21°C	Andrea et al. (2013)

(Continued)

TABLE 4.1 (*Continued*)
SRB Involved in the Treatment of Sulfate and Metal Containing Wastewater

Species	Carbon Source	Bioreactor	HRT	Temp.	Reference
Desulfovibrio-Desulfomicrobium, Desulfotomaculum, Desulfococcus-Desulfonema-Desulfosarcina	Lactate and phthalate	UASB	-	37°C	Dar et al. (2005)
Desulfovibrio, Desulforucans	Sodium Lactate	Flat-plate flow reactors	-	Room temp	Beyenal and Lewandowski (2004)
Desulfotomaculumkuznetsovii, M.thermoautotrophica, Mb. Thermoautotrophicus	Methanol	Batch reactor	-	60°C	Goorissen et al. (2004)
Desulfovibrio, Clostridium, Citrobacter and Cronobacter genera	Wine waste	UASB	9 days	21°C±1°C	Assunção Martins et al. (2011)
Desulfomicrobium	Formate	Fluidized bed reactor	24 hours	9°C	Auvinen et al. (2009)

The dissimilatory sulfate reduction comprises multistep chemical reactions, in which sulfate is activated by adenosine triphosphate (ATP). After transportation of sulfate into cell, ATP is consumed to produce adenosine 5'-phosphosulfate (APS). APS is then reduced to sulfite and adenosine monophospahte (AMP) by the APS reductase followed by conversion of sulfite reductase to sulfide. The pathway of dissimilatory sulfate reduction is shown in Figure 4.1.

4.3 ELECTRON DONORS FOR SULFATE REDUCTION

Acidic metal and sulfate-containing wastewater possesses less concentration of chemical oxygen demand (COD) and shows less sulfate removal efficiency. Hence, external carbon source is required by the sulfate-reducing bacteria to oxidize sulfate to elemental sulfur. There are varieties of carbon source or electron donors are available to complete the metabolism process of SRB. Theoretically 0.67 mole of carbon is required to degrade 1 mol of sulfate. An electron donor plays a very crucial role in biological sulfate reduction process. Compare to oxygen and nitrate, sulfate is considered as less favorable electron acceptor, so that additional carbon source as a supplement of electrons are provided to sulfate-reducing bacteria. SRB utilize this carbon source to oxidize sulfate to end products and gain energy for the growth and maintenance. The metabolism process of SRB using electron donor is expressed in following reaction.

$$2CH_2O + SO_4^{2-} \rightarrow 2HCO^{3-} + H_2S$$

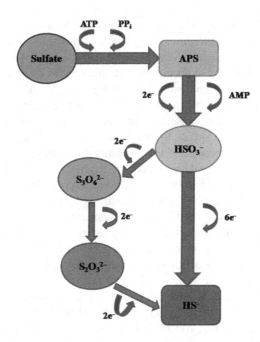

FIGURE 4.1 Pathways of dissimilatory sulfate reduction.

Here, CH_2O represents the organic carbon source that is oxidized by sulfate-reducing microorganism to generate bicarbonate and sulfides. The produced bicarbonate neutralizes acidity and favors the precipitation of metals.

The electron donor required by SRB can be of any type of carbon material which is easily available and cost effective. Generally, simple alcohols (ethanol, methanol, butanol), volatile fatty acids (propionate, lactate, butyrate, acetate, formate), pure H_2, carbon monoxide (CO) or mixture of synthesis gas is considered as a source of electron donor. Hydrogen is one of the most demanding electron donor due to its free energy for sulfate reduction (Weijima et al., 2002). On the other hand, in the economic point of view, pure hydrogen is not cost effective. To overcome from this situation researchers are using CO as cost effective alternative for H_2. Meanwhile, high concentration of CO is toxic for microorganisms. In this context, mixture of synthesis gas (H_2+CO) is used for efficient sulfate reduction. Now days, mixture of organic waste such as rice husk, animal manures, leaf munch, woodchips, sawdust, composted manure are also used as a efficient carbon source. Organic waste is attractive electron donor due to ease of availability and non-toxic also. Spent mushroom compost is having paramount nutrients and dissolved organic carbon. It is easily available and cost effective also. In passive system of metal and sulfate-containing wastewater known as successive alkalinity producing system (SAPS), mushroom compost is widely used for creation of organic layer (Das paper). Molasses is very rich in carbon content as it comprises mainly sugar and it is easily available as a waste product from sugar producing industry. Wine waste is also considered as one of the efficient source for carbon and electron donor as this contains mainly ethanol

(Costa et al., 2009). Wine waste as carbon source can be used by some specific species of sulfate-reducing bacterium at very limited conditions (Martins et al., 2009). Few literatures reported algae as an attractive electron donor for sulfate reduction. Algal biomass is considered as promising carbon source as it contains exopolysaccharides to work as an electron donor (Ayala-Parra et al., 2016).

The choice of electron donor is one of key factor to determine the performance of sulfidogenic bioreactor because it corresponds to the ability of sulfate-reducing microorganism to convert sulfate to sulfide. Hao et al. (2013) have reviewed variety of electron donor in context of their benefits and drawbacks. They pointed out the various genus of sulfate-reducing bacteria based on specific electron donor and electron acceptor. The different types of carbon source having substantial effect on the performance of bioreactor in context to starting up period, cost, availability and conversion of sulfate to their end products (Liamleam & Annachhatre, 2007). The selection of electron donor is to be considered as very important step in the biological sulfate reduction of sulfate-bearing wastewater. The following points should be considered in selection of electron donor/carbon source: (i) The ability of electron donor to completely remove sulfate and metal from wastewater without producing additional toxic pollutants (Liamleam & Annachhatre, 2007); (ii) The minimum cost and availability; (iii) The electron donor should be feasible for sulfate-reducing bacteria for their metabolism; (iv) Thermodynamic and kinetic parameters consideration to ensure the rate of reaction. The knowledge of energy of bacterial metabolism and growth will provide better understanding of predominate bacterial performance including thermodynamics. The Gibbs energy providing energy of reaction. Table 4.2 summarizes biochemical reaction occurred based on different carbon sources and their corresponding free energy.

Based on the data presented in Table 4.2, pyruvate has the most energetic reaction; therefore, it is the most favorable reaction for bacteria, and could be the best electron donor based on energy. The most frequently used electron donor is discussed in the following sections.

TABLE 4.2
Free Energy for Sulfate Reduction Reactions with Different Electron Donors

Reactions	ΔH°	$\Delta H^\circ/H_2$	ΔG°	$\Delta G^\circ/H_2$
$4\,H_2 + SO_4^{2-} \rightarrow S^{2-} + 4H_2O$	−196.46	−48.91	−123.98	−30.93
$4\,\text{Methanol} \rightarrow 3CH_4 + HCO3^- + H2O + H^+$	-	-	−316	-
$CH_4 + SO_4^{2-} \rightarrow CO_2 + H2O + S^{2-}$	40.55	10.11	16.72	4.18
$2\,\text{Ethanol} + SO_4^{2-} \rightarrow 2\,\text{acetic acid} + 2H_2O + S^{2-}$	−55.59	−13.92	−59.36	−20.06
$4\,\text{Formate} + SO_4^{2-} \rightarrow 4HCO3^- + S^{2-}$	−213.6	−53.5	−182.67	−45.56
$4\,\text{Pyruvate} + SO_4^{2-} \rightarrow 4\,\text{Acetate} + 4\,CO_2 + S^{2-}$	−351.12	−87.78	−331.06	−82.76
$2\,\text{Lactate} + SO_4^{2-} \rightarrow 2\,\text{Acetate} + 2CO_2 + 2H_2O + S^{2-}$	−79.42	−19.85	−140.45	−35.11
$2\,\text{Malate} + SO_4^{2-} \rightarrow 2\,\text{Acetate} + 2CO_2 + 2\,HCO^{3-} + S^{2-}$	154.66	38.66	−181.00	−45.35
$\text{Glucose} + SO42^- \rightarrow HS^- + 2\text{Acetate}^- + 2HCO3^- + 3H^+$	-	-	−358.2	-
$\text{Acetate} + SO_4^{2-} \rightarrow CO_2 + HCO_3^- + S^{2-} + H_2O$	48.07	12.12	−12.41	−3.09

4.3.1 Lactate

Lactate is one of the organic substrate used for sulfate reduction. As a carbon source and electron donor, lactate is used by various species of sulfidogenic microorganism. The following reaction describes sulfate reduction to hydrogen sulfide by using lactate.

$$2C_3H_5O_3^- + SO_4^{2-} \rightarrow 2C_2H_3O_2^- + HS^- + 2HCO_3^- + H^+ \text{ (Oyekola et al., 2009)}$$

Fang et al. (2011), reported complete removal of sulfate by using sodium lactate as a source of electron donor and carbon source. Biodegradation of sulfate-containing synthetic wastewater has been done in laboratory scale continuous flow stirred tank reactor having composition of wastewater are Zn (47.1 mg/L), Cu (19.8 mg/L), Cr (5 mg/L), and SO_4^{2-} (4 g/L). Complete reduction of sulfate and metal removal was observed with sodium lactate as a electron donor as well as pH was increased from 5 to 8 which is required effluent characteristics. They showed complete removal efficiency (~100%) when COD/SO_4^{2-} ratio was 17 and removal efficiency reached (~15%) when COD/SO_4^{2-} was 4 g/L. Singh et al. (2011) presented removal of sulfate, Cr and COD from real effluent in a small scale bioreactor. Removal efficiency of Cr, COD and sulfate was identified on the basis of effect of different hydraulic retention time (HRT), initial metal concentrations, various carbon sources and temperatures. Experiment was carried out at initial concentration of 50 mg/L with various carbon sources as sodium lactate, glucose, sucrose and fructose. The temperature was varied as 30°C, 35°C, 37°C, and 40°C with 0–144 hours of hydraulic retention time. It was found that maximum Cr and sulfate removal was observed as 96% and 82%, respectively, with lactate as a effective carbon source and electron donor.

Brahmacharimayum and Ghosh (2014) was fabricated laboratory scale packed bed reactor to study the sulfate reduction. In this present work, sulfate reduction to elemental sulfur was carried out under different operating condition of HRT, COD/SO_4^{2-} ratio and SO_4^{2-} concentration by using sodium lactate as a carbon source or electron donor. The performance of packed bed reactor was analyzed by varying initial sulfate concentration from 1,000, 1,200, 1,400, 1,600, and 1,500 mg/L by keeping COD 1,000 mg/L to provide COD/SO_4^{2-} ratio of 1, 0.8, 0.7, 0.62, and 0.67. The hydraulic retention time was varied as 18, 20, 24, and 30 hours to study the effect of HRT on the performance of packed bed reactor. Results revealed that maximum SO_4^{2-} reduction of 97% observed with initial SO_4^{2-} concentration of 1,500 mg/L and COD/SO_4^{2-} ratio at 24 hours HRT when lactate was used as carbon source. Oyekola et al. (2010) investigated effect of feed sulfate concentration on the kinetics of anaerobic sulfate reduction using lactate as the carbon source and electron donor by a mixed SRB culture. The study was operated across range of residence time (0.5–5 days) and initial feed sulfate concentration (1–10 g/L).

Yuan et al. (2014) presented an integrated reactor system for simultaneous removal of COD, sulfate and ammonium from synthetically made food industry wastewater. The integrated system consisted of 4 units as sulfate reduction and organic carbon removal (SR-CR), autotrophic and heterotrophic denitrifying sulfide removal (A&H-DSR), sulfur reclamation (SR) and aerated nitrification (AN) for complete removal of carbon, sulfur and nitrogen (C-S-N) to their elemental form. Lactate was used as a source of electron donor in the integrated system and the effects of each operational parameters viz HRT, sulfide/nitrate ratio, reflux ratio between A and H-DSR and AN units and loading rates of COD, sulfate and ammonium on the formation of elemental sulfur were investigated. It was found that lactate showed greater flexibility to convert sulfate to elemental sulfur with maximum removal of 98%COD, 98% sulfate and 78% nitrogen. Zhao et al. (2010) investigated the effect of different carbon sources on the starting up duration of sulfidogenic bioreactor. The different carbon sources utilized by sulfate-reducing bacteria in this study was molasses, acetate/ethanol, lactate, and glucose, which was grown on same seed sludge. The selection of suitable carbon source is required for the acceleration of startup period which is directly related to high efficiency sulfate reduction and production of organic end products (Acetate, propionate, and butyrate). It was found by this study that lactate fed bioreactor showed 99% sulfate reduction within very short time of operation as 15 days compared to other mentioned carbon sources. However, acetate and ethanol fed bioreactor proposed longer time to obtain steady state condition which is due to seize relatively less energy.

$$2CH_3CHOHCOO^- + SO_4^{2-} \rightarrow 2CH_3COO^- + 2HCO_3^- + HS^- + H^+ \Delta G^\circ = -160 \ kJ(1)$$

$$CH_3COO^- + SO_4^{2-} \rightarrow 2HCO^{3-} + HS^- \Delta G^\circ = -48 \ kJ(2)$$

$$2CH_3CH_2OH + SO_4^{2-} \rightarrow 2CH_3COO^- + HS^- + H^+$$

Verma et al. (2015) reported the removal of Cr (VI) from synthetic solution by using mixed culture of mesophilic sulfate-reducing bacteria. Sodium lactate was used as carbon source and electron donor in the batch biosorption experiment by cultured SRB to remove Cr (VI), COD and sulfate. Operational parameters such as pH, HRT, temp, and initial Cr (VI) was optimized by performing batch experiment in serum bottles. The present study concluded that 89.2% Cr (VI), 81.9% COD and 95.3% sulfate reduction was achieved at 50 mg/L initial Cr (VI) loading and sodium lactate as carbon source for 7 days of HRT.

4.3.2 ORGANIC WASTE

Acid mine drainage and another effluents comprising sulfate-bearing wastewater shows deficiency in sufficient carbon source to degrade sulfur components and metals present in wastewater. Hence, external carbon source is required to complete the biodegradation mechanism. Organic waste possesses high amount of carbon source to achieve required biological performance. There are variety of organic waste available to work as a source of carbon and electron donor such as mixture of wood chips, sawdust, cheese way, wine waste, spent mushroom compost, rice husk, coconut husk chips, animal manures, sewage sludge etc. Sufficient amount of organic matter shows efficient reduction of sulfate-bearing wastewater. Several researchers has done work using organic waste as a carbon source to reduce sulfate and metal concentration from waste water.

Kijjanapanich et al. (2014) developed in situ remediation technology to reduce the gypsum content of mine soils as well as huge amount of sulfate with the help of sulfate-reducing bacteria. The mine soils effluent contains high amount of sulfate as 150 g/kg. Experiment was conducted by using cheap organic substrates having low or no cost. The organic substrate such as rice husk, pig farm wastewater treatment sludge and coconut husk chips were mixed in ratio (60:20:20) by volume to provide electron donor for sulfate-reducing bacteria. The highest removal of sulfate (59%) was obtained when the soil was mixed with 40% of organic substrate mixture. The gypsum content of the soil reduced from 25% to 7.5% in context with the sulfate reduction. Hence, this study proves that mixture of organic substrates (rice husk, pig farm wastewater treatment sludge, and coconut husk chips) can be effectively utilized as a source of electron donor for sulfate-reducing bacteria to reduce sulfate content from gypesiferous soils.

Das et al. (2012) explored work on chemobioreactor for the continuous treatment of acid mine drainage. The method used in the treatment is SAPS by using spent mushroom compost (SMC) as electron donor for sulfate and metal reduction. In the SAPS system, spent mushroom compost is mostly used for organic layer formation. Spent mushroom compost comprises of nitrogen, phosphorus, potassium and dissolved organic carbon (DOC). These nutrients required can be used as a trace metals for the growth of sulfidogenic bacteria.The growth of sulfate-reducing bacteria supports due to conversion of polysaccharide content of spent mushroom compost to fatty acids and alcohols by hydrolytic fermentative anaerobic bacteria. 77% iron and 90% sulfate removal has been observed when DOC of spent mushroom compost was 50 mg/L. Sudden drop in sulfate removal (31%) has been observed when DOC was less than 50 mg/L. This is due to low availability of DOC which corresponds to slow growth of SRB. Results revealed that SRB population belongs to the availability of DOC. So that, spent mushroom compost having capability for the growth of SRB as well as absorption of metals removal too.

Wu et al. (2010) presented study based on in situ bioremediation of AMD using rice straw as sole substrate. Simultaneously wood chips was also used in same experiment to check the comparative ability as a carbon source. Generally acid mine drainage is having pH in range of 3–5. It was observed from experiment that treatment of AMD by using rice straw as a substrate increased sulfate removal rate from 8.67 to 21.77 mg/L-day with change in pH observed from 1 to 7. At the same time, treatment of AMD with the help of wood ships as a substrate showed rate of sulfate removal from 3.80 to 11.95 mg/L-day. Temperature Salso having very crucial role on the sulfate removal efficiency. As it has been investigated in present study that sulfate removal rate decreased from 13.93 to 9.91 mg/L-day by using rice straw as a substrate when temperature reduced from 25°C to 5°C whereas decline rate of sulfate removal has been observed from 7.43 to 4.98 mg/L-day with woodchips as a substrate. Rice straw and woodchips both comprising sufficient organic carbon for SRB to metabolize sulfate to sulfide, although previous literature showed that rice straw is having more hemicellulose and solvent extractable carbon content compare than woodchips (Chang et al., 2000). Although rice straw is easily available than woodchips and it is more powerful to degrade sulfate in vary critical condition as low pH and temperature.

Choudhary and Sheoran (2011) reported comparative study by using organic waste and cellulosic waste as substrate for sulfate and metal removal. In this study, babool & mango wood chips, sawdust were considered as cellulosic waste and cow, buffalo & goat manure, sugarcane waste, pearl & proso millet were considered as organic waste. Results revealed that mixture of buffalo, cow & goat manure increased pH from 2.75 to 7.10, whereas cellulosic waste increased pH from 4.83–5.3, which is not considered as required effluent characteristics. Maximum possible removal of metals observed with manure as substrates as 99.3% Fe, 99.9% Cu, 99.8% Zn, 99.1% for both Ni and Co and 73.8% Mn within 10 days of maximum retention period. It was found that cellulosic waste required more time (>10 days) to degrade the sulfate and mixture of metals. Literatures reported that cellulosic waste required extra supplement of easily incorporative electron donor such as sucrose, dextrose, pyruvate, cheese way, lactate, ethanol, or methanol etc. Whereas mixture of manures doesn't require any additional carbon source for biodegradation. Hence, mixture of animal manures can be used as effective carbon source.

Costa et al. (2014) investigated biodegradation of arsenic and sulfate with the help of sulfate-reducing bacteria capable growing on solid support as chicken feathers. Powdered chicken feathers are low cost and easily available organic waste having very rich nutrient content. Present study reported that powdered chicken feathers are capable to degrade sulfate and metal containing effluent with combination of sodium lactate as additional carbon source. It was found that arsenic removal increased from 38% (presence of sodium lactate only) to 80% (combination of sodium lactate & powdered chicken feathers). Similarly, less sulfate removal (13.4%) was observed when only powdered chicken feathers was carbon source and sulfate removal increased (27%) when combination of powdered chicken feathers and sodium lactate was used. So, it can be concluded that chicken feathers can also be used as effective carbon source for biodegradation. Sahinkaya et al. (2009) found waste sludge as efficient carbon source for biological treatment of sulfate and metal containing wastewater. The present study showed the suitability of waste sludge as a carbon and seed source to work at low temperature (8°C) for microbial sulfate reduction. To achieve this, two different sludge source viz waste activated sludge and anaerobically digested sludge was taken to compare their ethanol and acetate oxidation ability. It was found that 68% sulfate removal was obtained with waste sludge as single substrate whereas only 16% removal achieved with anaerobic digested sludge. After completion of experiment results revealed that waste sludge having potential for microbial degradation of sulfate-containing wastewater at low temperature.

4.3.3 ETHANOL

Ethanol is considered as one of the efficient electron donor or carbon source for sulfate-reducing bacteria due to ease of availability and low cost. It can be used at large scale applications. Previous literatures reported ethanol is more efficient than lactate in context to metal sulfide precipitation. Under sulfidogenic condition, ethanol is converted to acetate which is further oxidized to form carbon dioxide. Here acetate production is rate limiting step because during long period of operation possibility of acetate accumulation may occur which is ultimately affects sulfate reduction and sulfide precipitation in bioreactor. Following reaction involved in conversion of ethanol to carbon monoxide through SRB.

$$2CH_3CH_2OH + SO_4^{2-} \rightarrow 2CH_3COO^- + HS^- + H^+ + 2H_2O \qquad (4.1)$$

$$CH_3COO^- + SO_4^{2-} \rightarrow 2HCO_3^- + HS^- \qquad (4.2)$$

Kousi et al. (2011) demonstrated a batch upflow fixed-bed sulfate-reducing bioreactor for the treatment of synthetic wastewater containing divalent solution of iron, zinc, copper, nickel and sulfate using ethanol as sole substrate. Synthetic wastewater having concentration of divalent metal ions are 100 and 200 mg/L while sulfate are 1,700 and 2,130 mg/L, respectively. COD is supplied as ethanol and sulfate is supplied as sodium sulfate in theoretical stiochiometric ethanol/sulfate ratio. Results revealed that 85% sulfate removal obtained with effluent pH 3.5–7. This is due to complete oxidation of ethanol to acetate by sulfate-reducing bacteria. Hence, ethanol can be used as dominant electron donor for biological reduction.

Sahinkaya and Yucesoy (2010) did experiment on four stage anaerobic baffled reactor (ABR) for treatment of acidic sulfate and metal containing wastewater at 35°C for 250 days of period of operation. Ethanol was implemented in bioreactor for sulfidogenic bacteria as a source of carbon source/electron donor. It was found that 70%–92% sulfate reduction, 80%–94% COD oxidation and more than 99% metal precipitation observed. The alkalinity reached to 7 from 3 after sulfidogenic ethanol oxidation. So, this study established that ethanol supplemented anaerobic baffled reactor having potential for the treatment of acidic and metal containing wastewater.

Zhao et al. (2010) investigated the effect of different carbon source on the starting up period of sulfidogenic bioreactor. The availability and complexity of carbon source corresponds to the starting up period of bioreactor. The four different carbon source (molasses, glucose, acetate/ethanol, glucose) were supplemented to compare their capability in the shorter starting up of bioreactor. Rasool et al. (2015) reported anaerobic treatment of sulfate-rich synthetic textile wastewater by using three sulfidogenic sequential batch reactors. Simultaneously, experiment was performed to compare the effect of three different co-substrates viz lactate, glucose and ethanol on sulfate reduction and dye degradation. It was observed from result that reduction in co-substrate concentration eventually affects the color removal and sulfate removal efficiency. With reduction in co-substrate COD concentration from 3,000 to 500 mg/L, sulfate reduced from 98.42% to 30.27%, whereas color removal efficiency were observed from 98.23% to 78.46%, respectively. The present study proved that ethanol and lactate as electron donor were improved the sulfate and color removal efficiency.

Ucar et al. (2011) presented study with three stage fluidized bed reactor system by using ethanol as an electron and carbon source for SRB at 35°C to sequentially precipitate Cu and Fe from synthetically made acid mine drainage. The system incorporated with two pre-settling tanks before sulfidogenic ethanol fed fluidized bed reactor for the precipitation of metals. Results demonstrated that three stage fluidized bed reactor can be used for efficient removal of sulfate and COD with more than 99% Cu and Fe precipitation. Rodriguez et al. (2012) was used a bench scale upflow anaerobic sludge blanket (UASB) reactor for the treatment of acid mine drainage with ethanol supplemented carbon source. COD/SO$_4^{2-}$ ratio plays very crucial role in biological sulfate reduction and in this study variations in COD/SO$_4^{2-}$ ratio was applied to investigate the reactor performance. It was found that an increase in COD/SO$_4^{2-}$ ratio from 0.67 to 1 considerably increased the COD removal efficiency from 51.4% to 67.4% as well as sulfate removal efficiency increased from 46.3% to 85.6%. Thus UASB reactor with ethanol as external carbon source allows significant removal of COD and sulfate from wastewater.

4.3.4 SYNTHESIS GAS

Synthesis gas is generally a mixture of H$_2$, CO, and very often CO$_2$ also. Generally, synthesis gas is produced by thermal gasification of coal, biomass or other organic matter and steam reforming of natural gas. It is extensively used as a byproduct of coal burner, which is considered as a cheap alternative for H$_2$ (Parshina et al., 2010). Hydrogen, alone itself is very efficient carbon source for sulfate-reducing bacteria because of its free energy that make it more useful than methanogenic bacteria. Although H$_2$ is not cost effective as it is not easily available too. So that, H$_2$ with combination of CO as synthesis gas can be used as cheap alternative for electron donor.

Sipma et al. (2007) investigated a gas lift reactor for sulfate reduction by using CO as electron donor at 55°C with different hydraulic retention time. Inoculums used in this study was anaerobic sludge which is comprising thermophilic hydrogenogenic carbon monoxide converting microorganism to H2. It was observed that higher amount (up to 90%) of CO-derived H$_2$ consumption occurred by methanogens at HRT more than 9 hours which corresponds to the low sulfate reduction efficiency (<15%). However, 85% sulfate reduction obtained at HRTs <4 hours corresponds to the 95% consumption of the CO-derived H$_2$ by sulfate-reducing bacteria. Results concluded that sulfate reduction were controlled by amount of CO supplied as a carbon source and electron donor.

Sinharoy et al. (2015) successfully isolated hydrogenogenic carboxydotrophic bacteria having capability of sulfate reduction by utilizing anaerobic biomass from a large scale up flow anaerobic sludge blanket reactor. Anaerobic mixed bacterial consortia were collected from five different sources to examine their biological CO conversion efficiency. Reported results were showed that among five different sources, anaerobic granular sludge biomass presents maximum conversion of CO (85.62%) to divalent hydrogen and maximum sulfate reduction (50.65%). Touzé et al. (2008) designed a pilot fixed-bed column to treat sulfate and metal bearing wastewater. A fixed-bed was inoculated with a bacterial populations containing the sulfate-reducing organism known as *Desulfomicrobium norvegicum*. In this study, mixture of H$_2$ and CO$_2$ was used as carbon source. The pilot fixed-bed column was operated in continuous feeding conditions for 36 days which was decreased down to 8.5 hours. It was observed that temperature is having direct correlation with sulfate reduction as sulfate reduction increased from 35 to 95 mg/dm^3-h when temperature increased from 5°C to 17°C. To optimize the cost, optimization study has done on the basis of temperature regulations and input of carbon source.

Parshina et al. (2010) reviewed the current state of genomics of CO-oxidizing SRB with the utilization, inhibition characteristics and enzymology of CO metabolism. They have given detailed description of various genus responsible for sulfate reduction by using CO as electron donor. Isolation and characterization study was performed and after that comparative analysis of different types of geneus has done on the basis of their metabolism pathways and capability to reduce sulfate into end products. Sousa et al. (2015) fabricated hydrogen fed bioreactor operated at haloalkaline conditions to treat sulfate-rich streams. The present work was investigated based on sulfate reduction activity, microbial community and biomass aggregation. It was observed that low sulfate reduction volumetric rates obtained due to slow growth and lack of biomass aggregation. Thepresent study unable to show the mechanism behind biomass aggregation.

4.3.5 METHANOL

Methanol can be considered as a attractive electron donor for anaerobic thermophillic sulfate reduction because it is relatively cheap chemical and easily available. It is used successfully as electron donor in denitrification process (Weijma et al., 2002a,b) Methanol can be directly utilized by SRB or indirectly by involvement of other anaerobic microorganism (Liamleam & Annachhatre, 2007). Although, it was observed that methanol was not best among other electron donors as the reduction rate is very slow in the case of acidic mine waters. The reason might be due to at acidic pH, methanol also works as a alternative to organic acids in mesophillic temperature (*Geotechnolgy I*, 2014 springer). Moreover, their is a strong competition between methanogens, homoacetogens and sulfate-reducing organisms to use methanol best at mesophiilic temperature. It was observed that sulfate-reducing bacteria works better at thermophilllic temperature because at high temperature (50°C–65°C), they stimulated metalbollic activity and growth rates (Weijma and Stams, 2001). Desulfovibrio strains, D.africanus DSM 2603, D.salexigenss, Desulfotomaculum oreientis, Desulfobacterium catecholicum are found to use methanol as electron donor and carbon source (Hard et al., 1997).

Tsukamoto and Miller (1998) performed column study at room temperature to reactivate previously performed bioreactors by using variety of organic substrates (manure and wood chips) to decrease the sulfate. To increase the effectiveness of this systems, they further added lactate and methanol as carbon sources to a depleted cow manure substrate. It was observed that 69% sulfate reduction obtained at with lactate whereas 88% reduction obtained by using methanol as substrate. So, it was found that methanol can be utilized as a superior substrate for sulfate reduction.

Outstanding research has done by the Weijmain the sulfate reduction with the help of methanol as carbon source. He performed biodegradation experiment at all operating parameters (pH, mesophillic & thermophillic temp). Results demonstrated by Weijma et al. (2003) was the effect of sulfate on methanol conversion in mesophillic upflow anaerobic sludge bed reactors. It was found that more than 90% methanol was converted to methane, only 5%–10% was utilized for sulfate reduction. Conversion of methanol to their intermediate products by different organisms such as sulfate-reducing, methanogenic and homoacetogenicbacteria is represented in following reactions:

$$\left(\text{Sulfate reducing}\right) 4CH_3OH + 3SO_4^{2-} \Rightarrow 4HCO_3^- + 3HS^- + 4H_2O + H^+$$

$$\left(\text{Methanogenic}\right) 4CH_3OH \Rightarrow 3CH_4 + HCO_3^- + H_2O + H^+$$

$$\left(\text{homoacetogenic}\right) 4CH_3OH + 2HCO_3^- \Rightarrow 3CH_3COO^- + H^+ + 4H_2O$$

Methanol is converted to methane very efficiently in upflow sludge blanket reactor and present of sulfate also. So that, it is concluded by present study that methanol can not be used as suitable electron donor for sulfate-reducing bacteria under mesophillic conditions in upflow sludge blanket reactor.

4.4 SUMMARY AND CONCLUSION

Water coming from various industries containing high amount of metal and sulfate-rich wastewater. This wastewater can be treated by using biological sulfate reduction. For this, adequate amount of electron donor is required for complete removal. Different types of electron donor or carbon source is available which can be use by sulfate-reducing microorganism to achieve complete removal efficiency. While selecting electron donor, cost, removal efficiency, formation of intermediate products and availability should be considered. One can choose suitable electron donor based on study.

REFERENCES

Andrea, I. S., Stams, A. J. M., Amils, R., Sanz, J. L. (2013). Enrichment and isolation of acidophilic sulfate-reducing bacteria from Tinto-River basin. *Environmental Microbiology Reports, 5*, 672–678.

Assunção, A., Martins, M., Silva, G., Lucas, H., Coelho, M. R., & Costa, M. C. (2011). Bromate removal by anaerobic bacterial community: mechanism and phylogenetic characterization. *Journal of Hazardous Materials, 197*, 237–243.

Auvinen, H., Nevatalo, L. M., Kaksonen, A. H., & Puhakka, J. A. (2009). Low-temperature (9° C) AMD treatment in a sulfidogenic bioreactor dominated by a mesophilic *Desulfomicrobium* species. *Biotechnology and Bioengineering, 104*(4), 740–751.

Ayala-Parra, P., Sierra-Alvarez, R., & Field, J. A. (2016). Algae as an electron donor promoting sulfate reduction for the bioremediation of acid rock drainage. *Journal of Hazardous Materials, 317*, 335–343. Bernardes de Souja, C. M., Mansur, M. B.(2011). Modelling of acid mine drainage (AMD) in columns. *Brazilian Journal of Chemical Engineering, 28*, 425–432.

Bernardes de Souja, C. M., Mansur, M. B. (2011). Modelling of acid mine drainage (AMD) in columns. *Brazilian Journal of Chemical Engineering, 28*, 425–432.

Beyenal, H., Lewandowski, Z. (2004). Dynamics of lead immobilization in sulfate reducing biofilms. *Water Research, 38*, 2726–2736.

Brahmacharimayum, B., Ghosh, P. K. (2014). Sulfate bioreduction and elemental sulfur formation in a packed bed reactor. *Journal of Environmental Chemical Engineering, 2*, 1287–1293.

Burns, A. S., Pugh, C. W., Segid, Y. T., Behum, P. T., Lefticariu, L., Bender, K. S. (2012). Performance and microbial community dynamics of a sulfate-reducing bioreactor treating coal generated acid mine drainage. *Biodegradation, 23*, 415–429.

Cao, J., Zhang, G., Mao, Z., Li, Y., Fang, Z., Yang, C. (2012). Influence of electron donors on the growth and activity of sulfate-reducing bacteria. *International Journal of Mineral Processing, 106–109*, 58–64.

Castro, H. F., Williams, N. H., Ogram, A. (2000). Phylogeny of sulfatereducing bacteria. *FEMS MicrobiolEcol, 31*, 1–9.

Choudhary, R. P., Sheoran, A. S. (2011). Comparative study of cellulose waste versus organic waste as substrate in a sulfate reducing bioreactor. *Bioresource Technology, 102*, 4319–4324.

Costa, P. F., Matos, L. P., Leao, V. A., & Teixeira, M. C. (2014). Bioremoval of aresenite and sulfate-reducing capacity growing on powdered chicken feathers. *Journal of Environmental Chemical Engineering, 2*, 70–75.

Coulton, L. B., Bullen, C., Dolan, J., Hallet, C., Wright, J., Marsden, C. (2003). Wheal jane mine water active treatment plant-design, construction and operation. *Land ContamReclam, 11*, 245–52.

Dar, S. A., Kuenen, J. G., Muyzer, G. (2005). Nested PCR-Denaturing gradient gel electrophoresis approach to determine the diversity of sulfate reducing bacteria in complex microbial communities. *Applied and Environmental Microbiology, 71*, 2325–2330.

Dar, S. A., Yao, L., Dongen, U. V., Kuenen, J. G., Muyzer, G. (2007). Analysis of diversity and activity of sulfate reducing bacterial communities in sulfidogenic bioreactors using 16 r RNA and dsrB genes as molecular markers. *Applied and Environmental Microbiology, 73*, 594–604.

Das, B. K., Mandal, S. M., Bhattacharya, J. (2012). Understanding of the biocgemical events in a chemo-bioreactor during continuous acid mine drainage treatement. *Environmental Earth Sciences, 66*, 607–614.

Fang, D., Zhang, R., She, Z. (2011). Sulfate and heavy metals removal in a sulfate reducing bioreactor treating mildly acidic wastewater. *IEEE.*

Geurts, J. J., Sarneel, J. M., Willers, B. J., Roelofs, J. G., Verhoeven, J. T., & Lamers, L. P. (2009). Interacting effects of sulphate pollution, sulphide toxicity and eutrophication on vegetation development in fens: a mesocosm experiment. *Environmental Pollution, 157*(7), 2072–2081.

Gibson, G. R. (1990). Physiology and ecology of the sulphate-reducing bacteria-a review. *Journal of Applied Bacteriology, 69*, 769–797.

Glombitza, F. (2001). Treatment of acid lignite mine flooding water by means of microbial sulfate reduction. *Waste Management, 21*, 197–203.

Goorissen, H. P., Stams, A. J. M., Hansen, T. A. (2004). Methanol utilization in defined mixed cultures of thermophillic anaerobes in the presence of sulfate. *FEMS Microbiology Ecology, 49*, 489–494.

Hao, T. W., Wei, L., Lu, H., Chui, H. K., Mackey, H. R., van Loosdrecht, M. C. M., Chen, G. H. (2013). Characterization of sulfate-reducing granular sludge in the SANI(r) process. *Water Research, 47*, 7042–7052.

Hard, B. C., Friedrich, S., Babel, W. (1997). Bioremediation of acid mine water using facultativelymethylotrophic metal-tolerant sulfate-reducing bacteria. *Microbiology Research, 152*, 65–73.

Houten, B. H. G. V., Roset, K., Tzeneva, V. A., Dijkman, H., Smidt, H., Stams, A. J. M. (2006). Occurrence f methanogenesis during start-up of a full-scale synthesis gas-fed reactor treating sulfate and metal-rich wastewater. *Water Research, 40*, 553–560.

Icgen, B., Moosa, S., Harrison, S. T. L. (2007). A study of the relative dominance of selected anaerobic sulfate-reducing bacteria in a continuous bioreactor by fluorescence in situ hybridization. *Microbial Ecology, 53*, 43–52.

Johnson, D. B., Hallberg, K. B. (2005). Acid mine drainage remediation options: A review. *Science of Total Environment*, *338*, 3–14.

Jong, T., Parry, D. L. (2003). Removal of sulfate and heavy metals by sulfate reducing bacteria in short-term bench scale upflow anaerobic packed bed reactor runs. *Water Research*, *37*, 3379–3389.

Kieu, H. T. Q., Müller, E., Horn, H. (2011). Heavy metal removal in anaerobic semicontinuous stirred tank reactors by a consortium of sulfate-reducing bacteria. *Water Research*, *13*, 3863–3870.

Kijjanapanich, P., Annachhatre, A. P., Esposito, G., Lens, P. N. L. (2014). Use of organic substrates as electron donors for biological sulfate reduction in gyperiferous mine solis from Nakhon Si Thammarat (Thailand). *Chemosphere*, *101*, 1–7.

Klok, J. B. M., Van den Bosch, P. L. F., Buisman, C. J. N., Stams, A. J. M., Keesman, K. J., Janseen, A. J. H. (2012). Pathways of sulfide oxidation by haloalkaliphillic bacteria in limited-oxygen gas lift bioreactors. *Environmental Science and Technology*, *46*, 7581–7586.

Kousi, P., Remoundaki, E., Hatzikioseyian, A., Brunet, F. B., Joulian, C., Kousteni, V., Tsezos, M. (2011). Metal precipitation in an ethanol-fed sulfate-reducing bioreactor. *Journal of Hazardous Materials*, *189*, 677–684.

Liamleam, W., Annachhatre, A. P. (2007). Electron donors for biological sulfate reduction. *Biotechnology Advances*, *25*, 452–463.

Martins, M., Faleiro, M. L., Barros, R. J., Verissimo, A. R., Costa, M. C. (2009). Characterization and activity studies of highly heavy metal resistant sulphate-reducing bacteria to be used in acid mine drainage decontamination. *Journal of Hazardous Material*, *166*, 706–713.

Muyzer, G., Stams, A. J. M. (2008). The ecology of and biotechnology of sulphate-reducing bacteria. *Microbiology*, *6*, 441–454.

Oyekola, O. O., Van Hille, R. P., & Harrison, S. T. (2009). Study of anaerobic lactate metabolism under biosulfidogenic conditions. *Water Research*, *43*(14), 3345–3354.

Oyekola, O. O., Hille, R. P. V., Harrison, S. T. L. (2010). Kinetic analysis of biological sulfate reduction using lactate as carbon source and electron donor. *Chemical Engineering Science*, *65*, 4771–4781.

Parshina, S. N., Sipma, J., Henstra, A. M., Stams, A. J. M. (2010). Carbon monoxide as an electron donor for the biological reduction of sulfate. *International Journal of Microbiology*, *2010*, 319527.

Rasool, K., Mahmoud, K. A., Lee, D. S. (2015). Influence of co-substrate on textile wastewater treatment and microbial community changes in the anaerobic biological sulfate reduction process. *Journal of Hazardous Materials*, *290*, 453–461.

Rodriguez, R. P., Oliveira, G. H. D., Raimundi, I. M., Zaiat, M. (2012). Assessment of a UASB reactor for the removal of sulfate from acid mine water. *International Biodeterioration & Biodegradation*, *74*, 48–53.

Sahinkaya, E. (2009). Microbial sulfate reduction at low (8°C) temperature using waste sludge as a carbon and seed source. *International Biodeterioration& Biodegradation*, *63*, 245–251.

Sahinkaya, E., Yucesoy, Z. (2010). Biotreatement of acidic zinc and copper-containing wastewater using ethanol-fed sulfidogenic anaerobic baffled reactor. *Bioprocess BiosystEng*, *33*, 989–997.

Sarti, A., Pozzi, E., Chinalia, F. A., Ono, A., Foresti, E. (2010). Microbial processes and bacterial populations associated to anaerobic treatment of sulfate-rich wastewater. *Process Biochemistry*, *45*, 164–170.

Silva, A. J., Varesche, M. B., Foresti, E., Zaiat, M. (2002). Sulfate removal from industrial wastewater using a packed-bed anaerobic reactor. *Process Biochemistry*, *37*, 927–935.

Singh, R., Kumar, A., Kirrolia, A., Kumar, R., Yadav, N., Bishnoi, N. R., Lohchab, R. K. (2011). Removal of sulphate, COD and Cr(VI) in simulated and real wastewater by sulphate reducing bacteria enrichment in small bioreactor and FTIR stydy. *Bioresource Technology*, *102*, 677–682.

Sinharoy, A., Manikandan, N. A., Pakshirajan, K. (2015). A novel biological sulfate reduction method using hydrogenogeniccarboxydotrophicmesophillic bacteria. *Bioresource Technology*, *192*, 494–500.

Sipma, J., Osuna, M. B., Lettinga, G., Stams, A. J. M., Lens, P. N. L. (2007). Effect of hyrauliic retention time on sulfate reduction in a carbon monoxide fed therophillic gas lift reactor. *Water Research*, *41*, 1995–2003.

Sousa, J. A. B., Plugge, C. M., Stams, A. J. M., Bijmans, M. F. M. (2015). Sulfate reduction in a hydrogen fed bioreactor operated at haloalkaline conditions. *Water Research*, *68*, 67–76.

Touzé, S., Battaglia-Brunet, F., Ignatiadis, I. (2008). Technical and economical assessment and extrapolation of a 200-dm3 pilot bioreactor for reduction of sulfate and metals in Acid Mine Waters. *Water, Air, & Soil Pollution*, *187*, 15–29.

Tsukamoto, T. K., Miller, G. C. (1998). Methanol as a carbon source for microbiological treatment of acid mine drainage. *Water Research*, *33*, 1365–1370.

Ucar, D., Bekmezci, O. K., Kaksonen, A. H., Sahinkya, E. (2011). Sequential precipitation of Cu and Fe using a three-stage sulfidogenic fluidized-bed reactor system. *Minerals Engineering*, *24*, 1100–1105.

Verma, A., Dua, R., Singh, A., Bishnoi, N. R. (2015). Biogenic sulfides for sequestration of Cr(VI), COD and sulfate from synthetic wastewater. *Water Science*, *29*, 19–25.

Wang, J., Shi, M., Lu, H., Wu, D., Shao, M. F., Zhang, T., Ekama, G. A., van Loosdrecht, M. C. M., Chen, G. H. (2011). Microbial community of sulfate-reducing up-flow sludge bed in the SANI(r) process for saline sewage treatment. *Application Microbiology Biotechnology*, *90*, 2015–2025.

Weijima, J., Bots, E. A. A., Tandlinger, G., Stams, A. J. M., Hulshoff, P. L. W., Lettinga, G. (2002a). Optimization of sulfate reduction in a methanol-fed thermophillic bioreactor. *Water Resources*, *36*, 1825–1833.

Weijma, J., Stams, A. J. M., Hulshoff Pol, L. W., Lettinga, G. (2002b).Thermophillic sulfate reduction and methanogenesis with methanol in a high rate anaerobic reactor. *Biotechnology and Bioengineering*, *67*, 354–63.

Weijma, J., Chi, T. M., Hulshoff Pol, L. W., Stams, A. J. M., Lettinga, G. (2003). The effect of sulfate on methanol conversion in mesophillicupflow anaerobic sludge bed reactors. *Process Biochemistry*, *38*, 1259–1266.

Weijma, J., Stams, A. J. (2001). Methanol conversion in high-rate anaerobic reactors. *Water Science and Technology*, *44*, 7–14.

Weijma, J., Stams, A. J. M., Hulshoff Pol, L. W., Lettinga, G. (2002b). Thermophillic sulfate reduction and methanogenesis with methanol in a high rate anaerobic reactor. *Biotechnology and Bioengineering*, *67*, 354–63.

Wu, J., Lu, J., Chen, T., He, Z., Su, Y., Jin, X., Yao, X. (2010). In situ biotreatment of acidic mine drainage using straw as sole substrate. *Environmental Earth Sciences*, *60*, 421–429.

Yuan, Y., Chen, C., Liang, B., Huang, C., Zhao, Y., Xu, X., Tan, W., Zhou, X., Gao, S., Sun, D., Lee, D., Zhou, J., Wang, A. (2014). Fine tuning key parameters of an integrated reactor system for the simultaneous removal of COD, sulfate and ammonium and elemental sulfur reclamation. *Journal of Hazardous Materials*, *260*, 56–67.

Zhao, Y. G., Wang, A. J., Ren, N. Q. (2010). Effect of carbon sources on sulfidogenic bacterial communities during the starting-up of acidogenic sulfate-reducing bioreactors. *Bioresource Technology*, *101*, 2959–2969.

5 Bioremediation
A Prospective Tool for the Refurbishment of the Petroleum-Contaminated Areas

Manisha Goswami, Netra Prova Baruah, and Arundhuti Devi

5.1 INTRODUCTION

Substantial industrialization and large-scale urbanization have led to the production of huge volumes of various types of pollutants. The petroleum industry is one of the significant and fastest-growing sectors on which the economic growth of a nation depends (Varjani et al., 2020). Crude oil is processed and converted into 2,500 plus refined products by the petroleum refinery industry, predominantly gasoline, kerosene, and liquefied petroleum gas (LPG), distillate fuels, residual fuels, coke and asphalt, solvents, petrochemicals, and lubricants. Petroleum refining begins with distillation, followed by subjecting the converted products to various treatment and separation processes like hydro-treating, extraction, desalting, sweetening and cooling systems where an extensive volume of water is used. The distinctive characteristics and quantity of wastewater produced rests on the configuration or outline of the process. The wastewater generated in the refineries primarily consists of the following pollutants—oil, grease, suspended solids, Biological Oxygen Demand (BOD), Chemical Oxygen Demand (level of COD approximately 300–600 mg/L), sulfides, phenol (20–200 mg/L approximately), cyanide, 1–100 mg/L of benzene level, heavy metal (0.1–100 mg/L of chrome, 0.2–10 mg/L of lead), and a mixture of other contaminants. Petroleum wastewater contains a wide variety of pollutants such as petroleum hydrocarbons, heavy metals, oil and grease, suspended solids, phenols, sulfides, metal salts, etc. The conventional treatment of this industrial wastewater originating from petroleum refineries is established on the mechanical and typical physicochemical techniques such as separation of oil–water followed by coagulation and finally biological treatment. A light portion of aliphatic and aromatic hydrocarbons, and chlorinated organic compounds emanating from the industrial cooling liquids contaminate the petroleum wastewater received by the treatment plant. Among these pollutants that cause severe environmental menace, petroleum hydrocarbons (PHs) are of special category that needs urgent attention. Various activities

DOI: 10.1201/9781003368120-5

such as accidental spillage, pipeline leakage, transportation, industrial uses etc. may lead to the infiltration of PHs into the surrounding ecosystem (Khudur et al., 2018). Further it has been estimated that more than 1.5 million metric tons of crude oil enter the world's water bodies annually and among which 90% of it are consequences of human activities (Nikolopoulou et al., 2007). The problem is escalated further when these PHs reach the groundwater and contaminate it (Shahsavari et al., 2017).

PHs are known to be highly carcinogenic and attack essential organs like the liver and kidney (Sharma et al., 2020). PHs are mainly classified as aromatics, aliphatic, resins, and asphaltenes. The aliphatic hydrocarbons include alkanes, alkenes, alkynes, and cycloalkynes. Aromatic hydrocarbons may be monoaromatic or polyaromatic hydrocarbons (PAHs; Sarma and Prasad, 2015). PAHs consist of one or more fused benzene rings and in some hydrogen atoms in the benzene ring are replaced by ethyl or methyl groups to form new compounds like toluene, xylene, and ethylbenzene (Mohammadi et al., 2020). These compounds are highly toxic, mutagenic and carcinogenic, and not easily degradable in nature (Shahsavari et al., 2017). As the molecular weight of PHs increases, its toxicity also increases. Noteworthy that the toxicity of PHs of an organism is directly proportional to its bioavailability (Kuppusamy et al., 2020a). Therefore it is an utmost necessity of the present mankind to develop treatment techniques for the removal of PHs so that their harmful effects can be controlled.

Various treatment strategies have been developed for removing PHs from soil and water which includes bioremediation, membrane technology, electrochemical analysis, photocatalytic degradation, advanced oxidation, etc. (Haq and Kalamdhad, 2023; Haq et al., 2016a,b, 2017, 2022; Haq and Kalamdhad, 2021; Haq and Raj, 2018;). Of all these techniques, bioremediation of polluted soil and water is widely applicable because of its low cost and easy-to-operate methods. Thus, this chapter is aimed to illustrate the various impacts of PHs on the living organisms and to illustrate the various treatment techniques available for bioremediation of hydrocarbon-contaminated soil and water (Li et al., 2020; Dell'Anno et al., 2021).

5.2 EVALUATION OF BIOREMEDIATION EFFICIENCY

To assess the potency of the treatment analysis, many factors should be evaluated during any approach to bioremediation. Distinct techniques can be employed for this resolution which can be on the basis of biological, chemical, and/or physical indices of the investigation:

1. Physicochemical Approaches:
    a. Gas Chromatography-Flame Ionization Detection (GC-FID)
    b. Gas Chromatography-Mass Spectrometry (GC-MS)
    c. Infrared Spectroscopy (IR)
    d. Fluorescence Spectroscopy
    e. Biological Approaches
2. Culture-Dependent Techniques:
    a. Microbial Isolation and Enumeration: The preliminary isolation and calculation of the sum total microbial consortia and the degraders of

hydrocarbon contribute pivotal data about the biological depiction of contaminated soil.

b.   Biolog™ Plates: These plates are used to determine the metabolic activity of isolated microorganisms in polluted samples of the environment and Biolog Corporation was the developer of this technology in the late 1980s.

3. Culture-Independent Techniques:

a.   Enzyme Activity: The action of vital microbial enzymes is a widely used criterion in assessing PHC biodegradation and dehydrogenase enzymes are considered as an approximate of the oxidation capacity of microbes as they are in control of organic matter oxidation in the preponderance of microbial consortia in soil.

b.   Soil Respiration Tests: These are used to measure the oxygen consumption and CO_2 production during complete aerobic mineralization of PHC.

c.   Biomarkers: They are employed as indicators to assess PHC biodegradation. They are mainly of two types: Internal Petroleum Biomarkers and Cellular Biomarkers.

d.   Ecotoxicity Tests: To evaluate the toxicity of PHC-polluted samples, an acute toxicity test is the most frequently used approach. The main objective of the microbial ecotoxicology test, which is an advanced technique, is to precisely study the ecotoxicity outcomes of environmental pollutants on the entire population which can be attained by using heterogenous approaches together like analytical, molecular, enzymatic, and even toxicity.

4. Molecular-Based Techniques:

a.   Real-Time Polymerase Chain Reaction (qPCR): Real-time PCR or Quantitative PCR(qPCR) is a technique that provides real-time quantitative information related to a particular gene or data regarding a specific sequence of nucleic acid by relying on detection of the PCR products which are fluorescent-labeled and quantifying them in real-time during the period of PCR thermal cycle.

b.   Next-Generation Sequencing (NGS): NGS, a ground-breaking change in sequencing innovations, furnishes comprehensive statements related to the diversity and structure of microbial populations, the inherent contributions of organisms within the section and also the interactivity among the individuals within their populace.

5. Mixed Contamination: The efficient biodegradation of PHC is limited by the existence of many other additional co-contaminants which creates extreme complications during the process of bioremediation. Biostimulation, bioaugmentation, electrokinetics (EK), and bioelectrokinetics (BEK) are receiving wide acceptance in the field of activation and acceleration of the conventional degradation potentials.

5.3 BIOREMEDIATION TECHNIQUES FOR THE REMOVAL OF PHS CONTAMINATED SOIL AND WATER

Remediation techniques involve the application of various chemical, physical, and biological means. Among all these methods, bioremediation, that is, application of biological agents for breaking down and naturalization of various contaminants are widely used. Bioremediation is widely applicable because of its environment friendly and cost-effective nature (Iranzo et al., 2001; Kumar and Yadav, 2018). Various bioremediation techniques include bioaugmentation, biostimulation, bioventing, biosparging, bioslurping, biopiles land farming, windrows, natural attenuation, phytoremediation, etc. and the details of some of which are discussed below (Firooz Hosseini et al., 2023; Figure 5.1).

5.3.1 BIOAUGMENTATION

Bioaugmentation is one of the most efficient bioremediation techniques employed in the removal of PHs. In this technique, nonindigenous microorganisms having high degradation capacity are inoculated into the contaminated soil (Wu et al., 2013; Dadrasnia et al., 2018). Microorganism's selection used for bioaugmentation however is based on certain criteria (Szulc et al., 2014) such as

- Microorganisms must have the ability to degrade targeted contaminants.
- They must be highly adaptable to the existing environment.
- Microorganisms selected should not include pathogenic agents that may attack regional organisms.

However, various biotic and abiotic factors also affect bioaugmentation which includes moisture, pH, temperature, organic matter, aeration, nutrient content and the type of soil (Cho et al., 2000; Bento et al., 2005; Agnieszka and Zofia Piotrowska-Seget, 2010).

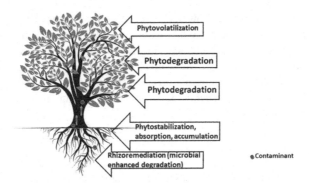

FIGURE 5.1   An overview of different bioremediation techniques.

Thus, the technique of bioaugmentation in general is the utilization of microorganisms either isolated from a contaminated site or genetically modified to degrade PHs polluted sites only after confirming that the native microorganisms are unable to degrade the PHs of the contaminated sites (Kumar and Yadav, 2018).

5.3.2 BIOSTIMULATION

The increase in the population of native microorganisms by the addition of nutrients artificially in the contaminated site is known as biostimulation. If the concentration of electron acceptors and nutrients is sufficient enough then only native microorganisms will grow sufficiently to degrade various harmful compounds (Dadrasnia et al., 2018). Places contaminated with hydrocarbons lack in the availability of organic nutrients such as phosphorous and nitrogen, which can be enhanced by the addition of NPK fertilizers, urea phosphate, phosphate and ammonium salts (Leahy and Colwell, 1990; Kumar and Yadav, 2018). Availability of essential elements like oxygen, phosphorous, carbon, nitrogen and hydrogen are required by the microorganisms for building up of macromolecules, thereby enhancing biodegradation. However far better results are obtained when inorganic nutrients are added in combination rather than one type of nutrient at one time (Sutherland et al., 2000).

5.3.3 BIOVENTING

Bioventing is the process of providing oxygen to the microorganisms living in the soil for stimulate the natural biodegradation capacity of the microorganisms to enhance soil quality. Low air-flow rates are generally used to supply enough amount of oxygen, which encourages the activity of microbes present in saturated zones (Kumar and Yadav, 2018). Bioventing reduces varieties of petrochemical compounds which include gasoline, fuel oil, bitumen, etc. However, there are some limiting factors that may affect the effectiveness of the techniques:

* Soil must be permeable. Low permeability reduces bioventing.
* Gases released at the surface of the soil must be supervised thoroughly.
* There must not be air present near the area of target because it may build up vapors around the air-injecting wells.
* Presence of chlorinated compounds may hamper biodegradation process.
* Low soil moisture developed due to bioventing may also hinders biodegradation.

5.3.4 BIOREACTOR

A bioreactor is fundamentally an engineered technique where biochemical conversion of materials is upgraded by promoting optimization of microbial activity or by in vitro action of cellular constituents of the cells of microbes (enzymes). The soil that is contaminated with oil is treated in these bioreactors which contain the system of aqueous slurry and this slurry-based bioreactor is considered as one of the foremost of high-speed total petroleum hydrocarbons (TPHs) bioremediation techniques because here TPH is easily transported to the cells of the microbes. There are other attractive and acceptable alternatives to the above-mentioned slurry bioreactors like the rotating drum bioreactors which can handle soils rich in PHs.

5.3.5 BIOSPARGING

This technique involves the injection of atmospheric air below the groundwater table or saturated zone for stimulating the degradation activity of the microorganisms toward organic pollutants (Kumar and Yadav, 2018; Chibuike and Obiora, 2014). The injection of nutrients and oxygen into the zone of saturation and application of pressure to proliferate the concentration of groundwater oxygen to revitalize the biological actions of the indigenous microbes for degradation of contaminants. This technique helps in saving energy. Small channels for the passage of air to reach the soil in the unsaturated zones are prepared by injecting air into the aquifers. Thus, as a result of biosparging, volatile pollutants are transferred to the unsaturated zone of soil, and soil vapor extraction is usually used to extract the volatile vapor and finally treat them at the surface (Held and Dorr, 2000). It enables minimization of contaminant concentration which is adsorbed by the soil, within the capillary perimeter exceeding the water table, and dissolved pollutants present in the groundwater. The noteworthy advantageous features of biosparging technology are:

1. Ease of installation of the equipment
2. Minimal inconvenience to site functioning
3. An average of 99.8% ammonia is mitigated in the whole process
4. The nitrates formed in the phase of nitrification are nearly completely removed by maintaining an acceptable dosage of organic carbon
5. No need to excavate or remove soil
6. Air inoculation rate is low minimizing the potential requirement for capturing vapor
7. Cost-competitive treatment

Limitations of Biosparging:

1. Biosparging depends on the elevated rate of airflow to attain contaminant volatilization and stimulate degradation. For this purpose, prediction of air-flow direction is essential and biosparging is unable to achieve this crucial target.
2. It is site or location-specific and can cause relocation of contaminants.

This process can be used to remediate ground water contaminated with benzene, toluene, ethylbenzene and xylene (BTEX). It is particularly beneficial in treating aquifers polluted with petroleum products, primarily diesel and kerosene. Its operation methodology is comparable with that of a closely resembling technique called in situ air sparging (IAS) which depends on powerful rates of airflow to effectuate contaminant volatilization, but biosparging on the other hand boosts biodegradation.

5.3.6 BIOPILES

When the piles of contaminated soil are mixed with nutrients and water, and the microbial communities present are stimulated by aeration, the piles are called biopiles (Kumar and Yadav, 2018; Chibuike and Obiora, 2014). It is usually applied for treatment of surface soil polluted with PHs. It is mostly applicable to the treatment of sandy granules-rich soils (Iturbe and Lopez, 2015). The main components required for operating this technique are temperature, nutrients, and aeration. It has been reported that through this technique remediation of large amount of contaminated soil can be done in limited space as compared to other bioremediation techniques. This green technology primarily targets at minimization of concentration of hydrocarbon which is existing in contaminated soil by application of the method of biodegradation. The constituents of this technique include: irrigation, systems collecting leachate along with nutrients, and a much essential bed for treatment. With proper maintenance of some parameters like aeration, supply of nutrients and temperature, this economical and constructive technology facilitates effectual biodegradation. It is specifically useful in pollution remediation in severe conditions of the environment like remarkably cold climate. The association of the objective of biopile and thermal system enables rise of microbial activities and pollutant availability thereby leading to the reduction in amount of time for treating the contaminated site and ultimately, increasing the biodegradation rate (Li et al., 2019).

Intensified bioremediation can be enabled by inoculating biopile with the high temperature warm air. The application of stationary biopiles in comparison to windrow technique in bioremediation of PHC-contaminated soils has been found to be less effort exhaustive, neither requiring any unconventional soil-turning supplies nor any associated on-site personnel and the employment of the former strategy resulted in uninterrupted, speedy mineralization. Humid biopiles possess very little concentration of TPH making them more preferable to heated and passive biopiles because conditions like ideal precipitation content, minimized leaching and minimum amount of volatilization of nominal degradable contaminants predominate. It has been earlier reported that this method enables bioremediation of a huge volume of polluted soils in a confined amount of area. To carry out experiments in laboratory is quite manageable and uncomplicated compared to execution of similar activities in land or practical field scale, but the design of biopile set-up can be easily scaled up to a field experimentation arrangement enabling achievement of interchangeable performance with those of laboratory researches (Martínez-Cuesta et al., 2023).

There are two unavoidable factors that need execution prior to processing the biopile technology—aeration and sieving of the contaminated soils to ensure its perceptible productivity. Certain bulking or volume-enhancing materials like barks of trees, environment friendly organic substances, sawdust, and straw have been incorporated to augment the process of remediation in a biopile arrangement. This eco-friendly technology conserves space in comparison to other ex situ biological remediation processes like land farming. But there are certain demerits of biopile set-up as well: maintenance expenditure and a dearth of supply of electricity in inaccessible regions which otherwise facilitates relentless air allocation in contaminated soil piles with the help of air pumps. Excessive heating of air creates drying of soil thereby restricting necessary activities of soil microorganisms and promoting vola-tilization instead of expected biodegradation and this ultimately delimits the overall process of bioremediation.

A technique for biodegrading hydrocarbons in soils is ecopiles, derived from biopiles—addition of phytoremediation to biostimulation with fertilization using nitrogen and employing indigenous bacteria to perform bioaugmentation. Ecopiling is a bioremediation technique that is ex situ in nature comprising of bioaugmen-tation, biopiling, biostimulation and phytoremediation merged together to attain contaminant removal from sediments as well as soils in a much effective and user-convenient manner. In a study conducted by Cuesta et al. (2023), at Carlow, Ireland, the approach of ecopiling was engaged to bioremediate local polluted soil by instal-lation of seven ecopiles in May 2019 to ameliorate 13,000 tons of industrially defiled soil. The actions of refinery plant of sugar-beet over a period of 80 years (1926–2005) resulted in hydrocarbon pollution in the surrounding locality. In the plantation of rye-grass (*Lolium perenne*) mixed with white clover (*Trifolium repens*) in those ecopiles, injection of 1.1228 L/m^3 in an average of previously segregated bacterial consortia (106–107 CFU/mL) from the polluted soils was performed. The cultured bacterial consortia were found to be rich in the following conspicuous genera: Acinetobacter, Extensimonas, Fulvimonas, and Pseudomonas as well as the dominant Proteobacteria, Actinobacteria, and Lysobacter were reported in the manuscript. The succession of the bacterial species, and evolving fungal communities to express restoration of soil, together with hydrocarbon (aliphatic and aromatic) monitoring in these ecopiles was thoroughly analyzed (Martínez-Cuesta et al., 2023). The results after the successful completion of the experiment have been found as follows:

- During a time period of 18 months of treatment, a substantial decrease by more than 90% of TPH levels on average and most of the TPH elimination took place during the initial treatment months.
- The reduction in soil pollution commanded an abstruse impact on the soil's microbial colonies.
- The progress of bioremediation resulted in an observation of microbial bio-diversity which was discernible for bacterial as well as fungal communities.
- A succession from the communities of microorganisms that colonize in con-taminated soils to those that are conventional of clean soils was witnessed.
- Successful hydrocarbon degradation, restoration and enrichment of soil quality.

In another study conducted by Oualha et al. (2019; Qatar), the employment of biopile technology was demonstrated together with the integration of bioaugmentation and biostimulation processes. They applied the biopile mechanism to minimize the content of hydrocarbon through the process of bioremediation. Their primary ambition was an acceleration of the aforementioned processes to execute and accomplish eviction of a particular portion of weathered hydrocarbons in curtailed time compared to typical unprocessed remediation. The quantification of the eliminated hydrocarbon fraction was implemented to reckon the functioning of the bioaugmentation approach and bio-stimulation technique. There exists a concomitant subsistence of biotransformation and removal of hydrocarbons in the process of bioremediation which gives rise to a delay in the complete expulsion of toxic hydrocarbons (Oualha et al., 2019).

The pollution levels of Dukhan, Qatar are quite elevated with soaring carbon and hydrogen constituents illustrating approximately 23% of the oil-contaminated land and dried matter suffering from soil weathering. The process of soil weathering is emphasized by the severe weather conditions of Qatar and its soil features which is the underlying justification for failure of majority of the applications of bioremediation for decontaminating hydrocarbon-polluted areas. The unsuccessful attempt of biostimulation utilizing urea led to fruitful application of ammonium nitrate which possessed an effective scope of biodegradation with efficient usage and treatment using selected optimal conditions: temperature of 37°C, 0.12% tween 80 surfactant, 10% of moisture content and C/N/P = 100/10/1. The biodegradation of pollutants existing in soil was facilitated by bioaugmentation utilizing native *B. sonorensis D1*. The reports of Colony Forming Unit (CFU), Fourier transform infrared (FTIR) and Gas chromatography (mass spectrometry/flame ionization detector; GC-(MS/FID)) analyses were corroborative of the efficiency of the green treatments. The percentage removal of diesel range organics (DRO) and PAH of the weathered soil contaminated with oil attained 39.2% and 32.4% simultaneously after a time period of 160 days when ammonium nitrate was administered as a derivation of nitrogen. On the other hand, urea generated an upsurge in pH level to 9.55 thereby restraining the degradation of oil (Oualha et al., 2019).

Another study conducted in the year 2019 by Rui Li et al., used biopile remediation technology for treatment of contaminant polyfluoroalkyl substances (PFASs) coexisting with PHs.

The system of biopiles is essentially a hybrid of two well-known bioremediation techniques—composting and landfarming. The efficacious treatment of earth's crust or surface soil contaminated with PHC is made viable by constructing aerated and composted soil piles for hindering loss of contaminants by ways of leaching and unwanted volatilization, which is a reconstructed pattern of landfarming. It involves disposing TPH-polluted medium into a reserve or stockpile with controlled features or parameters like moisture content, supply of nutrients and level of oxygen to instigate native soil microbes which can consume PHs as sources of carbon and transform them into carbon dioxide and essential water. Nevertheless, it is a little more high-priced compared to landfarming besides considering its assets like reliability and proven bioremedial alternative for treatment of TPH (Chibuike and Obiora, 2014).

In the year 2004, Iturbe et al. noticed at the termination of biopiling that percentage of TPH removal attained a value of 85% taking 4,500 mg/kg as its initial concentration after a treatment time period of 66 days. Gomez and Sartaj in the year 2014 monitored an experiment of TPH elimination with construction and maintenance of biopiles in field-scale and subjected it to heterogeneous microbial consortia with supplementary addition of compost under cold climate for over 94 days and observed 74%–82% removal of TPH. The concept of ecopiling which is a combination of passive biopiling and phytoremediation systems, was introduced by Germaine et al. in the year 2015 where the TPH-affected soil was subjected to amelioration using chemical fertilizers and further inoculating with bacterial consortia having the TPH-degrading property. This ecopile construction was completed by sowing a phyto-cap of *L. perenne* or perennial ryegrass and *T. repens* or white clover. The final result after 2 years' time was quite satisfactory—TPH levels reached below detectable limits from an initial concentration of 1,613 mg/kg in the ecopiles having petroleum hydrocarbon impact (Kuppusamy et al., 2020a).

5.3.7 COMPOSTING

Composting is a significant bioremediation process to treat oil-contaminated environments. It is a very austere approach associating the mechanized amalgamation of polluted sediment or soil consisting of compost and hydrocarbonoclastic bacteria bound by aerobic and temperate conditions. In the year 2008, Sinha et al. conducted a study involving use of earthworms for bioremediation of PAH-contaminated soil through vermicomposting. He confirmed that this technique worked efficiently toward enhanced PAH elimination from highly polluted gas work sites in a period of 12 weeks where the total initial concentration of PAH was found to be 11,820 mg/kg and the selection of earthworms included a variety of *Eisenia euginae*, *E. fetida*, and *Perionyx excavates* differing in age and size.

5.3.8 WINDROWS

In the windrow method, polluted soils are piled up in a particular shape so that aeration of the pile is done at a definite time period in order to increase bioremediation activity carried out by enhancing the degradation capacity of the microorganisms present in polluted soil (Azubuike et al., 2016). Addition of water is also needed for achieving bioremediation. When compared with biopile treatment, windrow treatment showed higher hydrocarbon removal rates depending on soil types (Kumar and Yadav, 2018; Coulon et al., 2010). However, because of the periodic turning in windrow method, bioremediation of soil contaminated with volatiles cannot be remediated successfully by this technique.

5.3.9 LAND FARMING

Land farming is one of the easiest ex situ bioremediation techniques, which is low cost and needs minimum operational tools. In land farming, hydrocarbon-contaminated soil is spread out at a height of 0.3–1.0 m with the addition and mixing of nutrients (Chibuike and Obiora, 2014). In common although it is said to be an ex situ bioremediation technique, but at times depending on the site of treatment it can be said an in situ technique. Whenever, pollutants lie 1 m in height below the ground, evacuation is not required for bioremediation (Nikolopoulou et al., 2013). Addition of nutrients in the form of nitrogen, phosphorous, potassium, together with good irrigation and tillage of soil is required. Aeration is another requirement for the success of land farming techniques. However, land farming techniques also have some limitations (Kumar and Yadav, 2018):

1. Land farming requires large space.
2. Conditions such as moisture and temperature are uncontrollable.

3. Inorganic pollutants remain nondegraded by this technique.
4. When tillage of soil is done, it creates dust particles which produce air pollution.
5. Volatile pollutants are to be pretreated to avoid air pollution.

5.3.10 NATURAL ATTENUATION

According to USEPA (1995), when the concentration of pollutant in the polluted sites are reduced using natural processes, it is called natural attenuation. In this method, biodegradation is carried out by the indigenous microflora of the soil without the involvement of human beings. Rate of degradation is generally slow in this method, as the whole process depends on availability of hydrocarbon-degrading bacteria, contamination level, and nutrients available (Iwamoto and Nasu, 2001). In order to prevent the side effects of this technique to the environment, long-term monitoring is very essential (Catherine and Yong 2004). Cost of this technique is comparatively low in regard to other methods as the requirement of external force is negative. However, one of the major limitations of natural attenuation is that, in order to reach a particular decontamination level, it takes a longer period of time as there is an absence of external force to accelerate the process. Further, on investigation, it was reported that natural attenuation does not have the capacity to remediate polyaromatic hydrocarbon and to reduce the soil ecotoxicity levels (García-Delgado et al., 2015).

5.3.11 BIOSLURPING

In this methodology, biodegradation of petroleum hydrocarbon contaminants occurs anaerobically, during which fragments of bioventing along with vacuum-enhanced pumping are the integrated for removal of unconstrained products from the soil and groundwater. It is an in situ remediation technique that merges vacuum-intensified pumping, withdrawing of soil vapor and bioventing to execute groundwater and soil remediation through meandering contribution of atmospheric oxygen and excitation of contaminant biodegradation. It is outlined for unencumbered yield retrievals like light non-aqueous phase liquids or LNAPLs, thereby treating and restoring capillaries together with zones which are saturated as well as unsaturated. The terrestrial regions polluted with volatile and partly-volatile organic compounds can also be made free from PHC contamination with the help of this technology. It utilizes a "slurp" extending into the layer of free product, which withdraws liquids, which can be soil gas and free yields, from the layer in a mode congruous to that of a straw drawing liquid particle up from any container. Compared to the rest of bioremediation approaches, bioslurping is regarded as the most economical or cost-compatible method. Here are a few essential factors related to the procedure:

1. The process of bioremediation is enhanced by performing extraction of the freely available product.
2. This technique minimizes smearing of aquifers. This smearing occurs when the water table is low and LNAPL traverses deep into the aquifers as

a result of which the vertical elevation of contaminant gets increased due to the interaction between saturated soils and contaminant.

3. It maximizes the genuine in situ bioremediation operation of the soil's vadose zone.
4. Moisture abundance in soil reduces permeability of air and diminishes its power of oxygen transfer; decrease in moisture content lowers activity of soil microbes.
5. The sites having less temperature lowers the process of bioremediation.
6. Prior to performing discharge, it is essential to treat the extracted groundwater and bioslurper emissions.

5.3.12 PHYCOREMEDIATION

The term "phycoremediation" is derived from two words: *phyco* meaning "algae" and *remediation* meaning "to heal or bring back to natural state"; hence, we can define phycoremediation as "using algae in the bioremediation process" of unwanted or wastewaters. The practice of trying to remediate wastes or wastewater using algae as a biological medium can be traced back to early civilizations where rivers, seas, and other waterways were perceived serviceable as toilets where humans and other animals would directly discharge their detritus considering the waterways as natural systems for discarding wastes. The multiplication of microalgae was reinforced by nutrients from the degrading or debased wastes and this in turn yielded both water photo-oxygenation as well as food supply for the remaining living beings in the aquatic ecosystem (Gupta et al., 2019).

5.3.13 PHYTOREMEDIATION

Application of living green plants and associated microbes directly for in situ biodegradation of contaminants in surface water, soils, and groundwater and obstructing subsequent pollution is called phytoremediation. In this technique of bioremediation, application of plant or vegetation combined with related microorganisms serves to remediate and restore the lost fertility of the contaminated regions. The association of microbes deactivates or debases environmental pollutants which are considered detrimental to the living bodies. The principle behind this progressive methodology is to optimize the conditions of plant root system for the stimulation of native microorganisms which are competent for degrading target pollutants. In comparison to various ex situ bioremediation techniques, phytoremediation is cost-effective, has lots of aesthetic benefits and easy to use (Kamath et al., 2004; Kumar and Yadav, 2018). Certain plant species possess the ability to uptake the contaminants from growth media transported to their tissue and thrive in polluted soils. For phytoremediation, grasses and trees are commonly used, where grasses are used for remediation of total PHs and PAHs and trees commonly used for remediation of toluene, benzene, xylene and ethylbenzene (Dadrasnia et al., 2018; Cook and Hesterberg, 2013). A few examples of plants (grasses, legumes, crops and weeds) with individual tolerance limits effective in the field of phytoremediation of petroleum hydrocarbon (PHC) contaminated soils are alfalfa, guinea grass, maize, prairie grass, red clover, *Mariscus alternifolius*, rye-grass, *Sida rhombifolia*, signal grass, sorghum, sunflower, and white clover. However, plants used as phytoremediators must fulfill the following criteria:

- The phytoremediators must possess good root systems. Fibrous or tap root system are preferred depending on contaminant depth.
- Amount of biomass above ground must be less so that animal consumption is less.
- Contaminants are less toxic toward the selected plants.
- Growth rate, adaptability and survival rate of the plant in existing conditions.
- Time required by the plant to reach the required degradation level.

There are various mechanisms which work simultaneously for the success of phytore-mediation. Sometimes plants work as a filter to remove contaminants or sometimes plants break down or suppress the pollutants (Kumar and Yadav, 2018). Generally, there are six methods used in phytoremediation:

- Rhizofiltration: wherein roots of plants are used for adsorption/absorption of contaminants from water and soil.
- Phytostabilization: wherein contaminants are bound by the plants to decrease its bioavailability.
- Phytodegradation: wherein contaminants are stored and degraded in the plant tissues.
- Phytostimulation: wherein degradation of pollutants is initiated by soil microbes and plant rhizosphere association.
- Phytovolatilization: wherein pollutants are taken up by the plants, where they get converted into nontoxic volatile material and which release later into the atmosphere.
- Phytoextraction: where the pollutants are taken up by the roots of the plants from contamination site and stored it in their shoots.

The method of enhancement of degradation and TPH removal during contaminated groundwater and soil remediation with the help of plant systems is time-consuming but is the most inexpensive technique because treatment of large areas of contami-nated soils can be done in situ without the need for excavation. There is also a pos-sibility of food chain and vegetation contamination through the application of this technique if advisable caution is not adopted. It uses a biological process in con-junction with physical and chemical techniques like hydraulic control, modification, volatilization, uptake and rhizo degradation to perform the removal, degradation, transformation or even stabilization of TPHs within land and groundwater. In the year 2015, Germaine et al. introduced a novel method for performing remediation of TPH-affected soils at a large scale which is termed as ecopiling-combined phy-toremediation and passive biopiling system. Here, the quality of TPH-polluted land was improved with the help of chemical fertilizers and inoculated with bacterial consortia possessing the ability to degrade unwanted TPH followed by construction and creation of passive biopiles. The completion of ecopile was performed by sowing a phyto-cap of *L. perenne*—perennial ryegrass and *T. repens*—white clover. The levels of TPHs that was initially of 1,613 mg/kg concentration, were found to be below the limit of detection after an investigation of 2 years from the day of construction.

Some examples of TPH concentration-reducing plants include Sorghum or *Sorghum bicolor* and common flax or *Linum usitatissimum* commonly called linseed which minimized the concentration of TPHs in a site from initial value of 40,000 to 18,500 mg/kg, that is, reduction by a value of 9,500; *Caroxexigua, Panicum virgatum*, and *Tripsacum dactyloides* resulted in 70% loss of TPH after a growth of 1 year; in 60 days *Viciafaba* showed 47% TPH degradation. A multidisciplinary experiment conducted using *Sagittaria alterniflora* combined with inorganic fertilizer, microbial seeding and soil oxidant enhanced the biodegradation of oil in coastal salt marshy areas and exhibited greater oil degradation or better phytoremediation rate in the drained state compared to flooded condition at wetlands. In the year 2003, Vervaeke et al. organized an experiment using willow trees and found that after a period of 1.5 years of treating contaminated areas, approximately 57% of mineral oil level was minimized in sediment and the decline in root zone was quite prominent (79%). A field study for the remediation of contaminated land by plantation of *Carexstricta, T. dactyloides*, and *P. virgatum* for more than a year was carried out by Euliss et al. in 2008 and he noticed a 70% degradation of PAHs from the experimented polluted area. Similarly, another study conducted in the year 2009 by Hultgren et al. exhibited 80% pyrene and 100% phenanthrene degradation with the plantation of *Salix viminalis. Festucapratensis* inoculated with diazotrophic bacteria namely, *Azospirillum sp.* and *Pseudomonas stutzeri*, were utilized to study degradation of a fusion of diesel fuel and PAHs conducted by Galazka and his team (2012). The achievement of greater than 70% remediation was observed in case of phytoremediatores inoculated with diazotrophs. Cai et al. in 2016 used *Festuca arundinacea* as phytoremediator bioaugmented with a halotolerant microbial consortium to investigate remediation of 64% heavy crude oil-polluted saline soils in 30 days. The use of PGPR or plant growth-promoting rhizobacteria by Gurska and his group (2009) improved phytoremediation in a comprehensive greenhouse test confirming efficacious lowering of TPH from 130 to 50 g/kg over 3 years. Endophytic fungi—*Neotyphodium coenophialum* and *Neotyphodium uncinatum*, infecting two types of grasses namely, *Festuca arundinacea* and *Festu capratensis* expunged 65%–85% PAHs and TPHs after a treatment period of 7 months (Soleimani et al., 2010). Memarian and Ramamurthy in the year 2012 demonstrated the potency of tween 80 surfactant solution under requisite aerobic conditions in rhizodegradation of oil integrated with its advantage on native soil microorganisms.

In a recent study conducted by Abdallah et al. (2022), *Acacia seiberiana* Tausch—an indigenous leguminous plant of Sudan, was employed to test its capacity to remediate oil-contaminated soil with concentrations: 0.5%, 1%, 1.5%, and 2% (w/w) crude oil and improve the action of soil microbes. Leguminous plants have the congenital power to fix atmospheric nitrogen and do not have to compete with soil microbes or struggle for the limited nitrogen supply at oil-contaminated sites; instead, they can instigate related microbes by delivering nutrients to the rhizosphere. The results exhibited noteworthy accomplishment of phytoremediation as the crude oil concentration did not disturb the plant growth of the Sudanese plant taken for experimentation purposes and the parameters—bacterial count, degradation percentage and TPH measurement also support the satisfactory results obtained. The effect on the lengths of shoot and root of *A. sieberiana* Tausch was investigated and its tolerance to 2% crude oil concentration was found but the plant failed to endure stress till a period of 6 months. However, legumes like *Acacia* spp. displayed highly competent root length performance on such polluted sites compared to other plants like *Desmodium glabrum* and *Mimosa camporum*, grown on oil-contaminated soils which were studied earlier. The soil samples collected for examining the degradation percentage of crude oil along with growth and development performance of *A. sieberiana* Tausch under oil stress conditions were obtained from the rhizospheric region of these seedlings. The reported results exhibiting the benefits of *A. sieberiana* Tausch plantation were: remarkable capability in oil degradation enhancement, high count of viable microbes in contaminated soil and unique phytoremediation mechanism (Abdallah et al., 2023).

In another study by Eze et al. (2022), the synergistic interactions between the perennial leguminous plant—*Medicago sativa*, commonly known as alfalfa or lucerne (widely grown for hay, pasturage, silage and possesses tolerance to diesel fuel toxicity), and microbial strain —*Paraburkholderia tropica* (can colonize barley roots and boost plant maturing) were reported. The degradation percentages of *M. sativa* alone were 72%, *P. tropica* alone was 86% and the combined treatment of *M. sativa* alone + *P. tropica* resulted in 96% degradation or practically total degradation of the hydrocarbon contaminants. A proper examination of the consequences of this study would prove beneficial for biotechnological applications in agriculture and achieve several other environmental remediation objectives.

The advanced and green technique of bioremediation—phytoremediation, is subjected to greater consent of populace because of its negligible maintenance, reduced environment-devastating character, and also it does not lead to any secondary pollutant creation. There are a few factors which govern the productivity and mechanism of the process: category of contaminant, bioavailability of the contaminant, plant put to use for phytoremediation and features and quality of the soil (Ogwuche et al., 2023). Figure 5.2 demonstrates the mechanisms involved in the method of phytoremediation:

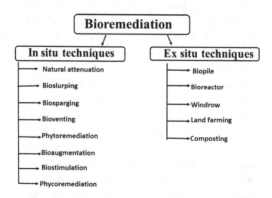

FIGURE 5.2   The mechanism involved in the process of phytoremediation.

5.4   GAPS AND LIMITATIONS

The techniques employed in bioremediation are favored over the remaining technologies because they are additionally inexpensive and receive superior public acceptance conventionally. Still there are certain unavoidable limitations that need consideration and must be taken into significance. Several review articles have taken these limitations into account and listed them in detail. The parameters affecting bioremediation have been put together by Boopathy (2000) regardless of them being regulatory, scientific and even non-scientific (Table 5.1).

The rate at which cells of microorganisms can convert pollutants is influenced by the process of mass transfer and hence is considered as a very important parameter. Boopathy (2000) mentioned that when mass transfer acts as a limiting factor, higher rates of biotransformation are not ensured by maximized capabilities of microbial conversion. Ultimately, the absence of mass transfer results in weathering or aging and bioavailability of contaminant is affected, that is, they their availability decreases with the passage of time.

TABLE 5.1
Technical Factors Restraining PHC Bioremediation

Factors	Limitations
Aerobic and anaerobic processes	Availability of acceptors of electron
	Insitu community of microbes
	Oxidation and Reduction potential
Bioavailability of pollutants	Equilibrium sorption
	Irreversible form of sorption
	Subsuming into humic matters
Environmental	Dearth of nutrients and superior substrates
	Repressed environmental conditions
Co-metabolism vs growth substrate	Alternate source of carbon
	Concentration and type of pollutants
	Interactions among microbes: their competition, predation and succession
Mass transfer	Diffusion and solubility of oxygen
	Miscibility or solubility with water
	Nutrient diffusion
Microbial	Enrichment of competent occupants
	Enzyme stimulation
	Horizontal gene delivery
	Generation of hazardous metabolites
Substrate	Chemical structure of contaminants
	Low concentration of pollutants
	Solubility, Toxicity

Source: Adaptation, Boopathy (2000).

Keeping apart the scientific features, there exists certain non-technical hindrances that need to be taken into consideration—potential to reach time limitations and agreeable regulatory discernment, liability, human resources and the ability to meet the target amidst others. The regulatory parameters command what pollutants need cleaning, to which range or dimension and determine the approaches to be taken into account.

Bioremediation is an innovative and prominently emanating industry and hence firm restrictions are inflicted on it. It needs the incorporation of several unavoidable and emerging disciplines like engineering, hydrogeology, microbiology, project management and soil science for achieving desirable objectives of projects but the paucity of specialized personnel is what deprives the accomplishment of successful proposals.

5.5 SUMMARY AND CONCLUSION

Much exploration has been done on the bioremediation of soil and water defiled or tainted with PHs and it has been considered as one of the most cost-effective, ecologically compatible, practically viable and productive technologies for healing the environment continuously undergoing laceration by certain unwanted human activities as well as ineluctable natural weathering processes. Bioremediation is an extensive and emerging field with various in situ and ex situ methodologies for the downright management of a diversified collection of environmental contaminants.

The aim of this chapter was to elucidate the various bioremediation techniques accepted and practiced for the total removal of PHs. The detoxification, removal, or degradation studies of PHCs and its co-contaminants from terrestrial and aquatic ecosystems utilizing plants, and microbial taxa like bacteria, algae and fungi effective in the domain of bioremediation and its genetic engineering have continuously contributed toward development of novel hybrid biological approaches. An endeavor needs to be made for generation of an interaction that is synergistic in nature among the environmental reverberation on the destiny and behavior of pollutants, agglomeration as well as interpretation of the most congenial bioremediation approach and another relevant technique that can maintain the proficient and accomplished maneuver and supervision of a bioremediation procedure. The redressal and restoration of a salubrious environment by expulsion of PHCs, one of the most relevant environmental foes of the recent decades, is a must to ensure a clean, green and sustainable environment. Furthermore, it can be concluded by saying that the accessibility and biodegradation of pollutants in any unadulterated or processed system and caliber of menace to the health of humans caused by diverse environmental contaminants would be effortlessly predicted by valuing multidisciplinary technologies.

REFERENCES

Abdallah, A. H., Elhussein, A. A., & Ibrahim, D. A. (2023). Phytoremediation of crude oil contaminated soil using Sudanese plant species *Acacia sieberiana* Tausch. *International Journal of Phytoremediation*, 25(3), 314–321.

Agnieszka, M., & Piotrowska-Seget, Z. (2010). Bioaugmentation as a strategy for cleaning up of soils contaminated with aromatic compounds. *Microbiological Research*, 165(5), 363–37.

Azubuike, C. C., Chikere, C. B., & Okpokwasili, G. C. (2016). Bioremediation techniques-classification based on site of application: Principles, advantages, limitations and prospects. *World Journal of Microbiology and Biotechnology, 32*, 1–18.

Bento, F. M., Camargo, F. A., Okeke, B. C., & Frankenberger, W. T. (2005). Comparative bioremediation of soils contaminated with diesel oil by natural attenuation, biostimulation and bioaugmentation. *Bioresource Technology, 96*(9), 1049–1055.

Boopathy, R. (2000). Factors limiting bioremediation technologies. *Bioresource Technology, 74*(1), 63–67.

Catherine, N. M., & Yong, R. N. (2004). Natural attenuation of contaminated soils. *Environmental International, 30*(4), 587–601.

Chibuike, G. U., & Obiora, S. C. (2014). Bioremediation of hydrocarbon-polluted soils for improved crop performance. *International Journal of Environmental Sciences, 4*(5), 840–858.

Cho, Y. G., Rhee, S. K., & Lee, S. T. (2000). Effect of soil moisture on bioremediation of chlorophenol-contaminated soil. *Biotechnology Letters, 22*, 915–919.

Cook, R. L., & Hesterberg, D. (2013). Comparison of trees and grasses for rhizoremediation of petroleum hydrocarbons. *International Journal of Phytoremediation, 15*(9), 844–860.

Coulon, F., Al Awadi, M., Cowie, W., Mardlin, D., Pollard, S., Cunningham, C., ... & Paton, G. I. (2010). When is a soil remediated? Comparison of biopiled and windrowed soils contaminated with bunker-fuel in a full-scale trial. *Environmental Pollution, 158*(10), 3032–3040.

Martínez-Cuesta, R., Conlon, R., Wang, M., Blanco-Romero, E., Durán, D., Redondo-Nieto, M., ... & Rivilla, R. (2023). Field scale biodegradation of total petroleum hydrocarbons and soil restoration by Ecopiles: microbiological analysis of the process. *Frontiers in Microbiology, 14*, 1158130.

Dadrasnia, A., Usman, M. M., Alinejad, T., Motesharezadeh, B., & Mousavi, S. M. (2018). Hydrocarbon degradation assessment: Biotechnical approaches involved. In *Microbial Action on Hydrocarbons*, 63–95. https://doi.org/10.1007/978-981-1 3-1840-5_4.

Dell'Anno, F., Rastelli, E., Sansone, C., Brunet, C., Ianora, A., & Dell'Anno, A. (2021). Bacteria, fungi and microalgae for the bioremediation of marine sediments contaminated by petroleum hydrocarbons in the omics era. *Microorganisms, 9*(8), 1695.

García-Delgado, C., Alfaro-Barta, I., & Eymar, E. (2015). Combination of biochar amendment and mycoremediation for polycyclic aromatic hydrocarbons immobilization and biodegradation in creosote-contaminated soil. *Journal of Hazardous Materials, 285*, 259–266.

Gupta, P. K., Ranjan, S., & Gupta, S. K. (2019). Phycoremediation of petroleum hydrocarbon-polluted sites: application, challenges, and future prospects. In *Application of Microalgae in Wastewater Treatment: Volume1: Domestic and Industrial Wastewater Treatment*, 145–162. https://doi.org/10.1007/978-3-030-13913-1_8.

Held, T., & Dörr, H. (2000) In situ remediation. *Biotechnology, 11*, 350–370.

Haq, I., & Kalamdhad, A. S. (2021). Phytotoxicity and cyto-genotoxicity evaluation of organic and inorganic pollutants containing petroleum refinery wastewater using plant bioassay. *Environmental Technology & Innovation, 23*, Article 101651.

Haq, I., & Kalamdhad, A. S. (2023). Enhanced biodegradation of toxic pollutants from paper industry wastewater using *Pseudomonas sp.* immobilized in composite biocarriers and its toxicity evaluation. *Bioresource Technology Reports, 24*, 101674.

Haq, I., Kalamdhad, A. S., & Pandey, A. (2022). Genotoxicity evaluation of paper industry wastewater prior and post-treatment with laccase producing *Pseudomonas putida* MTCC 7525. *Journal of Cleaner Production, 342*, Article 130981.

Haq, I., & Raj, A. (2018). Markandeya Biodegradation of Azure-B dye by *Serratia liquefaciens* and its validation by phytotoxicity, genotoxicity and cytotoxicity studies. *Chemosphere, 196*, 58–68.

Haq, I., Kumar, S., Kumari, V., Singh, S. K., & Raj, A. (2016a). Evaluation of bioremediation potentiality of ligninolytic *Serratia liquefaciens* for detoxification of pulp and paper mill effluent. *Journal of Hazardous Materials, 305*, 190–199.

Haq, I., Kumari, V., Kumar, S., Raj, A., Lohani, M., & Bhargava, R. N. (2016b). Evaluation of the phytotoxic and genotoxic potential of pulp and paper mill effluent using *Vigna radiata* and *Allium cepa. Advanced Biology, 2016*, 1–10, Article ID 8065736.

Iranzo, M., Sainz-Pardo, I., Boluda, R., Sanchez, J., & Mormeneo, S. (2001). The use of microorganisms in environmental remediation. *Annals of Microbiology, 51*(2), 135–144.

Iturbe, R., & López, J. (2015). Bioremediation for a soil contaminated with hydrocarbons. *Journal of Petroleum & Environmental Biotechnology, 6*(208), 2.

Iwamoto, T., & Nasu, M. (2001). Current bioremediation practice and perspective. *Journal of Bioscience and Bioengineering, 92*(1), 1–8.

Kamath, R., Rentz, J. A., Schnoor, J. L., & Alvarez, P. J. J. (2004). Phytoremediation of hydrocarbon-contaminated soils: Principles and applications. In *Studies in Surface Science and Catalysis* (Vol. 151, pp. 447–478). Elsevier. https://doi.org/10.1016/S0167-2991(04)80157-5.

Khudur, L. S., Shahsavari, E., Aburto-Medina, A., & Ball, A. S. (2018). A review on the bioremediation of petroleum hydrocarbons: Current state of the art. In *Microbial Action on Hydrocarbons* (pp. 643–667). https://doi.org/10.1007/978-981-13-1840-5_27.

Kumar, R., & Yadav, P. (2018). Novel and cost-effective technologies for hydrocarbon bioremediation. In *Microbial Action on Hydrocarbons* (pp. 543–565). https://doi.org/10.1007/978-981-13-1840-5_22.

Kuppusamy, S., Maddela, N. R., Megharaj, M., Venkateswarlu, K., Kuppusamy, S., Maddela, N. R., & Venkateswarlu, K. (2020a). An overview of total petroleum hydrocarbons. In *Total Petroleum Hydrocarbons: Environmental Fate, Toxicity, and Remediation*, 1–27. https://doi.org/10.1007/978-3-030-24035-6_1.

Kuppusamy, S., Maddela, N. R., Megharaj, M., Venkateswarlu, K., Kuppusamy, S., Maddela, N. R., & Venkateswarlu, K. (2020b). Ecological impacts of total petroleum hydrocarbons. In *Total Petroleum Hydrocarbons: Environmental Fate, Toxicity, and Remediation*, 95–138. https://doi.org/10.1007/978-3-030-24035-6_5.

Leahy, J. G., & Colwell, R. R. (1990). Microbial degradation of hydrocarbons in the environment. *Microbiological Reviews, 54*(3), 305–315.

Li, Q., Liu, J., & Gadd, G. M. (2020). Fungal bioremediation of soil co-contaminated with petroleum hydrocarbons and toxic metals. *Applied Microbiology and Biotechnology, 104*, 8999–9008.

Li, R., Munoz, G., Liu, Y., Sauvé, S., Ghoshal, S., & Liu, J. (2019). Transformation of novel polyfluoroalkyl substances (PFASs) as co-contaminants during biopile remediation of petroleum hydrocarbons. *Journal of Hazardous Materials, 362*, 140–147.

Martínez-Cuesta, R., Conlon, R., Wang, M., Blanco-Romero, E., Durán, D., Redondo-Nieto, M., ... & Rivilla, R. (2023). Field scale biodegradation of total petroleum hydrocarbons and soil restoration by Ecopiles: Microbiological analysis of the process. *Frontiers in Microbiology, 14*, 1277.

Mohammadi, L., Rahdar, A., Bazrafshan, E., Dahmardeh, H., Susan, M. A. B. H., & Kyzas, G. Z. (2020). Petroleum hydrocarbon removal from wastewaters: A review. *Processes, 8*(4), 447.

Nikolopoulou, M., Pasadakis, N., & Kalogerakis, K. (2007) Enhanced bioremediation of crude oil utilizing lipophilic fertilizers. *Desalination, 211*, 286–295.

Nikolopoulou, M., Pasadakis, N., Norf, H., & Kalogerakis, N. (2013). Enhanced ex situ bioremediation of crude oil contaminated beach sand by supplementation with nutrients and rhamnolipids. *Marine Pollution Bulletin, 77*(1–2), 37–44.

Ogwuche, C. E., Edjere, O., & Obi, G. (2023). Phytoremediation of crude oil polluted soil using dried yam peels (Discoreasp) as a case study in Niger Delta Environment of Southern Nigeria. *European Journal of Environment and Earth Sciences*, *4*(1), 1–5.

Oualha, M., Al-Kaabi, N., Al-Ghouti, M., & Zouari, N. (2019). Identification and overcome of limitations of weathered oil hydrocarbons bioremediation by an adapted *Bacillus sorensis* strain. *Journal of Environmental Management*, *250*, 109455.

Shahsavari, E., Poi, G., Aburto-Medina, A., Haleyur, N., & Ball, A. S. (2017). Bioremediation approaches for petroleum hydrocarbon-contaminated environments. In *Enhancing Cleanup of Environmental Pollutants*: *Volume 1*: *Biological Approaches*, 21–41. https://doi.org/10.1007/978-3-319-55426-6_3.

Sarma, H., & Prasad, M. N. V. (2015). Plant-microbe association-assisted removal of heavy metals and degradation of polycyclic aromatic hydrocarbons. *Petroleum geosciences: Indian contexts*, 219–236.

Sharma, K., Kalita, S., Sarma, N. S., & Devi, A. (2020). Treatment of crude oil contaminated wastewater via an electrochemical reaction. *RSC Advances*, *10*(4), 1925–1936.

Soleimani, M., Afyuni, M., Hajabbasi, M. A., Nourbakhsh, F., Sabzalian, M. R., & Christensen, J. H. (2010). Phytoremediation of an aged petroleum contaminated soil using endophyte infected and non-infected grasses. *Chemosphere*, *81*(9), 1084–1090.

Sutherland, T. D., Horne, I., Lacey, M. J., Harcourt, R. L., Russell, R. J., & Oakeshott, J. G. (2000). Enrichment of an endosulfan-degrading mixed bacterial culture. *Applied and Environmental Microbiology*, *66*(7), 2822–2828.

Szulc, A., Ambrożewicz, D., Sydow, M., Ławniczak, Ł., Piotrowska-Cyplik, A., Marecik, R., & Chrzanowski, Ł. (2014). The influence of bioaugmentation and biosurfactant addition on bioremediation efficiency of diesel-oil contaminated soil: feasibility during field studies. *Journal of Environmental Management*, *132*, 121–128.

United States Environmental Protection Agency (1995) Bioventing principles and practice vol.1: bioinventing principles. United States Environmental Protection Agency, Washington, DC.

Varjani, S., Joshi, R., Srivastava, V. K., Ngo, H. H., & Guo, W. (2020). Treatment of wastewater from petroleum industry: Current practices and perspectives. *Environmental Science and Pollution Research*, *27*, 27172–27180.

Wu, M., Chen, L., Tian, Y., Ding, Y., & Dick, W. A. (2013). Degradation of polycyclic aromatic hydrocarbons by microbial consortia enriched from three soils using two different culture media. *Environmental Pollution*, *178*, 152–158.

6 Perspective on the Treatment of Brewery Wastewater Using Microalga

Surbhi Sinha

6.1 INTRODUCTION

Brewery industries, although an important component of the country's economy, consume significant amounts of water throughout the manufacturing process and then discharge approximately 70% of it as effluent (Jaiyeola et al., 2016). Wastewater byproducts like yeast and spent grains, generated during the two primary steps of beer manufacturing (brewing and packing) are the main contributors to environmental contamination. Additionally, cleaning tanks, and machinery and flushing human waste also led to the infiltration of water sources. This effluent has significant levels of chemical oxygen demand (COD), nitrogen, phosphorus, and other organic loading, rendering it unfit for any beneficial application (Mata et al., 2012). After pretreatment, brewery wastewater may be released directly into (i) municipal sewers, (ii) nearby water bodies, or (iii) the brewery's wastewater treatment facility.

Partially treated brewery effluent released into water bodies poses environmental problems. The key environmental challenges raised by brewery operations include water use, wastewater, solid waste, byproduct generation, and energy use. This phenomenon causes environmental issues such as water scarcity, uncontrollable microbial growth that destroys aquatic lifeforms, and health issues in the communities near the discharge sites. Before discharging their wastewater into the environment, breweries must therefore cleanse and manage their effluents appropriately. Despite being widely employed in the treatment of brewery wastewater, conventional treatment procedures typically produce enormous volumes of sludge. Conventional methods of treatment also have high costs for operation and maintenance, which makes them even less economical. Furthermore, excessive chemical use may result in ecological imbalances (Niu et al., 2018; Singh et al., 2017).

The application of microalgae, an ecologically benign and cost-efficient water treatment technology, has been recognized as a solution to these challenges. Microalgae can effectively eliminate organic pollutants from wastewater and provide a valuable biomass byproduct (Khan et al., 2022). Wastewater treatment using algae and making use of its biomass is currently receiving interest on a global scale. This chapter examines the properties of brewery wastewater, existing treatment

DOI: 10.1201/9781003368120-6

technologies, as well as the utilization and possibilities of microalgae in brewery wastewater management. The chapter also explored the advantages and disadvantages of using microalgae treatment technologies to efficiently clean brewery wastewater for environmental protection.

6.2 GENERAL CHARACTERISTICS OF BREWERY WASTEWATER

Brewery wastewater has usually high moisture content. Because of the presence of organic components, this effluent often has a high COD and BOD concentration. Brewery effluent generally has a temperature range of 25°C–38°C. The pH values vary depending on the concentration and category of chemicals used in cleaning and sanitizing (e.g., caustic soda, phosphoric acid, nitric acid, etc.) wastewater. Nitrogen (N) and phosphorus (P) levels in wastewater are also affected by raw material handling and the amount of yeast present. The physicochemical characteristics of the brewery wastewater are listed in Table 6.1 (Subramaniyam et al., 2016).

6.3 CURRENT WASTEWATER TREATMENT APPROACHES

6.3.1 Pretreatment Method of Brewery Wastewater

Pretreatment is necessary for brewery wastewater due to its high TSS, TS, and turbidity, as well as its high content of organic and suspended particles (Tonhato Junior et al., 2019). The principal aim of the pretreatment approach for brewery wastewater is to modify the physical, chemical, and biological characteristics of the feed water. Physical, chemical, or biological approaches, or combinations of two or more of these, are utilized in the pretreatment procedure. However, the choice of pretreatment procedure is heavily influenced by the effluent's final discharge point. In a case, wherein, the brewery wastewater is not released into the municipal drainage system, simple primary and secondary treatments are needed. However, if the brewery wastewater is discharged into a municipal drain, pretreatment is mandatory to alleviate organic feeds on the municipal treatment plant and to comply with wastewater treatment regulations (Huang et al., 2009).

TABLE 6.1
Characteristics of Brewery Wastewater

pH	6–7
Phosphorous (mg/L)	18–20
Nitrogen (mg/L)	15
Manganese (mg/L)	0.06
Sulfur (mg/L)	17–19
Organic carbon (mg/L)	183
Sodium (mg/L)	238
Calcium (mg/L)	35

6.3.2 Physical Methods Utilized for Brewery Wastewater Treatment

The physical method of wastewater treatment has traditionally been utilized to decrease suspended matter from wastewater using gravity-force sedimentation (Ranjit et al., 2021). It also removes fat and oil from wastewater. Howbeit, the physical approach simply ousters suspended matter but does not decompose them. Screening, flow equalization mixing, flotation, and sedimentation are the physical procedures currently being utilized in wastewater treatment (Thakkar et al., 2006) system. Physical treatment often serves as the pretreatment stage in the brewery wastewater remediation.

6.3.3 Chemical Methods Utilized for Brewery Wastewater Treatment

The chemical approach requires pH modification or coagulation–flocculation of effluent by adding different chemicals to change its chemistry. Coagulation–flocculation is the primary phase in the chemical process for remediating wastewater (Bouchareb et al., 2021). Flocculation entails the agitation of chemically treated wastewater to produce coagulation, which enhances sedimentation performance by increasing particle size and thereby raising settling efficiency. Aluminum sulfate and ferric chloride are two examples of inorganic coagulants that have been widely used in wastewater treatment (Gao et al., 2023). Organic components in wastewater are oxidized during this treatment process by adding chemical compounds like chlorine, ozone-oxygen, or permanganate to produce CO_2, H_2O, and other non-offensive materials. Chemical flocculants are very effective but harmful to the environment and human health. To protect the microorganisms in wastewater, the pH needs to be maintained between 6 and 9. Due to the corrosive characteristics of H_2SO_4 and HCl as well as the discharge limitations imposed by sulfate and chloride, neutralizing the pH of wastewater by utilizing these chemicals is typically not suggested. However, waste CO_2 could be used as an acidifying agent to reduce the alkalinity (high pH) of wastewater before anaerobic digestion. Okolo et al. (2016) investigated the coagulation–flocculation performance of biocoagulant *Detarium microcarpum* seed powder (DMSP) by varying effluent pH, coagulant dosage, and settling time in the removal of suspended solid particles from brewery effluent (BE) at room temperature. The optimum pH was observed at 4.0, while 100.0 mg/L DMSP delivered optimum suspended solid particle removal. Menkiti et al. (2016) investigated the potential of chitin-derived coagulant (CDC) in the removal of suspended and dissolved particles (SDP) from BE via response surface methodology. Approximately 98% of dissolved particles were removed from the BE.

6.3.4 Biological Methods Utilized for Brewery Wastewater Treatment

BE necessitates effective treatment procedures that can degrade organic concentrations in wastewater. The wastewater is processed using both anaerobic (using aerobic bacteria) and aerobic (using activated sludge) digestion procedures to reduce organic content before discharge into the municipal sewer.

6.3.4.1 Anaerobic Approach

Anaerobic digestion is an organic process in which different types of microorganisms work together in the absence of oxygen to turn organic waste into biogas through a series of intermediates (Harirchi et al., 2022). This method of treatment is used a lot in breweries because the biogas that is made can be used to maintain the operating temperature or to make a profit. However, a variety of parameters, such as nutrients, organic loads, the carbon/nitrogen ratio, temperature, and pH of the wastewater, have an impact on anaerobic digestion (Polprasert, 2007). Anaerobic bacteria need the desired temperature to efficiently breakdown organic contaminants. However, most industrial anaerobic plants usually work in the mesophilic temperature range. On the other hand, the working bacteria are categorized according to their preferred pH range. Optimal pH ranges for acidogenic and methanogenic bacteria are, respectively, less than 6.0 and 7–8 (Sekine et al., 2022).

6.3.4.2 Aerobic Approach

The aerobic treatment uses aerobic microorganisms (bacteria) to digest organic materials in wastewater in the presence of oxygen. The process produces inorganic byproducts like carbon dioxide, ammonia, and water (Ahammad et al., 2020). Aerobic processes are more efficient in the digestion of pollutants. Microorganisms convert non-settleable sediments to settleable solids in the aerobic biological treatment process, followed by sedimentation, which allows the settleable solids to settle and separate. The following aerobic treatment methods are commonly employed in wastewater treatment: (i) activated sludge process, (ii) attached growth (biofilm) process, and (iii) the trickling filter technique. However, the selection of aerobic treatment methods depends heavily on the concentration of contaminants in the effluent. Due to the significant levels of organic contaminants in the effluent, trickling filters and activated sludge are typically utilized in the treatment of brewery wastewater.

Activated sludge methods are commonly utilized in wastewater treatment (Haq and Kalamdhad, 2021, 2023; Haq et al., 2016a, b, 2017, 2022; Haq and Raj, 2018). During this step of the process, the wastewater is directed into an activated sludge-primed, aerated, and agitated tank. Aeration devices vigorously mix the suspension of aerobic bacteria in the aeration tank, supplying oxygen to the biological suspension (Liu et al., 2019).

In the attached growth phase, aerobic biological activity produces a favorable environment for microorganisms to attach to the solid surface. During the trickling filter process, BE is sprayed on the surface of solid materials such as gravel and stone/plastics, which allows the effluent to trickle down via decomposing microbial media (Amenorfenyo et al., 2019). The use of both anaerobic and aerobic methods in wastewater treatment comes at a high capital expense. However, compared to aerobic treatment, the anaerobic procedure has a lower operational cost. Aerobic procedures are impeded by the physical and chemical fluctuation of the wastewater, the high cost of treatment, and the development of excessive sludge by microorganisms. Table 6.2 lists the comparison between aerobic and anaerobic water treatment systems.

TABLE 6.2

Comparison between Aerobic and Anaerobic Treatment Systems

Aerobic System	Anaerobic System
Aerobic wastewater treatment is a biological wastewater treatment process that uses an oxygen-rich environment	Anaerobic wastewater treatment is a process where anaerobic organisms break down organic material in an oxygen-absent environment
High cost	Relatively less costly
Requires more energy	Requires less energy
Aerobic wastewater treatment does not produce methane and carbon dioxide	Anaerobic wastewater treatment produces methane and carbon dioxide
Sludge yield is high	Sludge yield is relatively low
The activated sludge method, trickling filter, rotating biological reactors, and oxidation ditch arecexamples of aerobic wastewater treatment	Anaerobic lagoons, septic tanks, and anaerobic digesters are examples of anaerobic wastewater treatment

6.3.5 ADVANCED OXIDATION TREATMENT PROCESS

Advanced oxidation treatment techniques (AOPs) are commonly employed in the treatment of wastewater from distilleries and breweries. In this process, the initial stage of oxidation uses ozone, hydrogen peroxide, and UV radiation to form hydroxyl radicals (OH). The second phase involves the formation of precipitates by the reaction of organic loads with hydroxyl radicals (Sinha et al., 2022). AOP technologies are made possible by combining hydrogen peroxide/ultraviolet irradiation (H_2O_2/UV), zone/ultraviolet irradiation (O_2/UV), and ozone/hydrogen peroxide (O_2/H_2O_2). When dissolved into water, ozone reacts with a significant number of organic molecules, hence aiding in the removal of organic pollutants from wastewater. It interacts either directly as molecular ozone or indirectly by creating secondary oxidants in the form of hydroxyl radicals (•OH) and other free radical species. Another well-known AOP technique that utilizes the Fenton reaction is called Fenton's oxidation. This procedure involves the mixing of hydrogen peroxide with ion salts (Fe^{2+} or Fe^{3+}). The Fenton oxidation method generates hydroxyl radicals (•OH), which cause the precipitate formation and effluent decolorization. The Fenton process generates an environmentally favorable homogenous reaction (Raji et al., 2021). However, the commercial application of these procedures is still in the research and development phase. The use of AOP in the treatment of wastewater from breweries and other sources produced successful results and offers hope for the future of brewery wastewater management. However, the technique may necessitate additional treatment to reduce ozone, which may raise treatment costs. Also, turbidity and NO_3 provide challenges to the AOP procedures, which require attention.

6.3.6 ACTIVATED CARBON

Activated carbon is often used to clean municipal drinking water, industrial wastewater, and point-of-use and point-of-entry filters. The strong adsorption properties of activated carbon make it useful for removing various organic components from industrial wastewater (Lewoyehu et al., 2021). Carbon can be used to cleanse the whole flow of an effluent including various contaminants, or it can be utilized as a part of a multistage strategy to eliminate distinct contaminants present in the effluent. Fermentation and other brewing operations contribute to the effluent's foul stench. This effluent may contain some aromatic and carbon-sulfur-bonded compounds, which often have a poor odor and taste. These molecules attach to carbon particularly well. Additionally, because of its propensity to react with oxidizing substances like hypochlorous acid and chlorine dioxide, carbon may also be used as a dechlorination agent. In the brewing sector, activated carbons can be used as an odorless wastewater treatment solution. This method is more affordable and does not require energy or high water pressure.

6.3.7 BREWERY WASTEWATER TREATMENT USING MICROALGAE

Due to their capacity to take in nutrients and transform them into biomass, micro-algae are viewed as one of the beneficial wastewater treatment agents (Abdelfattah, et al., 2022). Nitrogen, phosphorus, and other nutrients in the wastewater are properly absorbed by microalgae for their growth throughout the brewery wastewater treatment process. Through their photosynthetic processes, microalgae freely release oxygen, which bacteria in wastewater use (He et al., 2022). Additionally, microalgae fix CO_2 by the assimilation of HCO_3 from CO_2 via respiration.

Lately, microalgae have been just used in the laboratory for the treatment of wastewater. Raceway ponds and photobioreactor technology have been used to treat microalgae effluent, including brewery wastewater. Raceway ponds have a shallow open system and are semi-circular at both ends. The system has paddle wheels that keep mixing the microalgae in the wastewater so that they can get sunlight and nutrients (Sawant et al., 2019). Photobioreactors are built in either vertical or horizontal columns. The design enables light to reach the microalgae. Microalgae are provided with sufficient CO_2 by the injection and circulation of carbon dioxide (Sirohi et al., 2022).

According to a study by Lutzu et al. (2016), *Scenedesmus dimorphus* can remove more than 99% of nitrogen (N) and phosphorous (P) from BE in just 1 week; nitrogen was lowered from 229 to 0.2 mg/L and phosphorus from 1.4–5.5 to 0.2 mg/L. Subramaniyam et al. (2016) cultured *Chlorella* sp. in BE and found that the microalgae eliminated total nitrogen, phosphorus, and organic carbon with remarkable growth. Similarly, Song et al. (2020) investigated the efficiency of four different microalgae strains (*Chlorella* sp. L166, *Chlorella* sp. UTEX1602, *Scenedesmus* sp. 336, and *Spirulina* sp. FACHB-439) for the treatment of hybrid brewery wastewater and CO_2 fixation. Experimental results indicated that *Chlorella* sp. UTEX1602 could tolerate 15% CO_2, and *Scenedesmus* sp. 336 presented good brewery wastewater purification performance for standalone CO_2 fixation and wastewater purification tests. Han et al. (2021) co-cultivated *Scenedesmus* sp. 336, *Chlorella sorokiniana* UTEX1602, and *Chlorella* sp. L166 achieves the target of simultaneous biomass accumulation, wastewater treatment, and biological carbon fixation. The experimental results showed that when the mixture ratio of *Scenedesmus* sp. 336 and *C. sorokiniana* UTEX1602 was set at 1:1 with the ventilation period of 6 h/d, the cell dry weight could reach 796.89 mg/L. Travieso et al. (2008) investigated the effectiveness of *C. vulgaris* in the remediation of the distillery. The authors determined that *C. vulgaris* eliminated over 98% of COD and BOD and that the final effluent may be safely discharged into the environment.

Although the microalgae approach may remove a lot of contaminants from BE, the technology has limitations when it comes to the removal of salt, odor, and color. For this technology to remove all contaminants from the effluent, it needs to be used with other cost-effective methods. For the polishing stage of wastewater treatment, microalgae-based technology can be paired with membrane technology. Microalgae require optimal light and temperature to grow. Due to the low sunlight and temperatures in temperate areas, this technology might not work there. Artificial lighting systems could be employed as an alternative; however, this could lead to a rise in treatment costs in these nations. However, biodiesel and other useful items can be made from biomass.

6.4 MICROALGAL PROSPECTS IN WASTEWATER TREATMENT

6.4.1 Cost-Efficiency

The growth of microalgae in wastewater can be maintained at a cost that is far lower than that of traditional wastewater treatment methods. Brewery wastewater contains organic loads that are conducive to the growth of microalgae, making it a very appealing option for environment-friendly and inexpensive wastewater treatment (He et al., 2022).

6.4.2 Minimal Energy Demand

Microalgae produce oxygen in the process of treating wastewater, and aerobic bacteria use this oxygen to break down the residual loads. When compared to the expense of mechanical energy for aeration in traditional wastewater treatment, this lowers the cost of energy. In the activated sludge process, 1 kg of BOD is removed using about 1 kWh of electricity. One kilogram of fossil carbon dioxide is released into the atmosphere throughout this procedure. Microalgae can treat 1 kg of BOD in brewery wastewater without any additional energy input and generate 1 kWh of electricity via methane generation (Kabir et al., 2022).

6.4.3 Abatement in Sludge Development

Every wastewater treatment facility has as its major purpose the reduction or eradication of sludge. A lot of chemicals are used in the traditional way of treating wastewater. Substantial chemical usage can lead to the production of sludge. This results in hazardous solid wastes that must be eliminated from the environment. Microalgae wastewater treatment does not require chemical additions, and sludge accumulates as algal biomass (Song et al., 2022).

6.4.4 EMISSION OF GREENHOUSE GASES

Global warming is a major source of concern around the world. Chemical and biological approaches have been taken as part of a coordinated effort to reduce CO_2, as stated by Wang et al. (2008). Separation, transport, and sequestration are all aspects of chemical methods. These methods are expensive and energy-intensive, thus there is a need for more affordable, sustainable methods to reduce the threat.

About 2,000,000 microalgae species can sequester CO_2, making them an emerging biotechnological method for CO_2 mitigation. The ability to fix CO_2 through photoautotrophic algae cultivation has the potential to reduce atmospheric CO_2. Due to their superior ability to absorb solar energy, microalgae develop at a far faster rate than their terrestrial counterparts. According to Li et al. (2008), microalgae have substantially better growth rates and CO_2 fixation capability than typical forestry, agricultural, and aquatic plants. Li et al. (2008) found that microalgae have significantly greater growth rates and CO_2 fixation capabilities than typical forest, agricultural, and aquatic plants. In comparison to conventional wastewater treatment methods, microalgae-based treatment eliminates tons of CO_2. In comparison to electromechanical treatment in a traditional oxidation pond, raceway ponds or high-rate algal pond (HRAP) systems eliminate 100–200 tons of CO_2 per ml of treated wastewater by utilizing bacteria, sunlight, and photosynthesis. Additionally, the uptake of nitrogen by algae may result in a further ton (100–200 mL) reduction in CO_2. As a result, microalgae-based wastewater treatment, or its integration into other wastewater treatment plants, can biologically reduce CO_2 levels. This approach is more affordable, efficient, and environmentally beneficial.

6.5 ADVANTAGES OF WASTEWATER TREATMENT USING MICROALGAE

6.5.1 AS BIO-FERTILIZERS

Excessive use of inorganic fertilizers modifies soil fertility by raising soil acidity and decreasing soil pH, respectively. Furthermore, inorganic fertilizers contain compounds such as nitrates and phosphates that are carried into water bodies by rain and sewage, potentially leading to eutrophication. Microalgae are useful in the agro-industry. The algal biomass captured from wastewater can be converted into plant fertilizers. These fertilizers enhance the mineral content and water retention capability of agricultural soils. Additionally, microalgae can fix nitrogen in the soil. *Scytonema* sp., *Nostoc* sp., *Aulosira* sp., *Toplythrix* sp., and *Plectonema* sp. all fix nitrogen in the soil and are usually used as bio-fertilizers (Zou et al., 2021).

6.5.2 AS ANIMAL FOOD

Aquaculture uses microalgae as live feed because of their high nutritional value and simple digestion (Kusmayadi et al., 2021). Microalgae are mostly made up of polysaccharides, cellulose, and starches, which make up 10%–57% of the dry matter. Harvested biomass can be used either directly or indirectly as a food source for larval bivalves, oysters, and shrimp. Li et al. (2021) produced selenium-enriched microalgae as a potential feed supplement in high-rate algae ponds treating domestic wastewater. Microalgae biomass can be used as a partial substitute in chicken feed for traditional proteins and carotenoids to improve the yellow color of broiler skin and egg yolk. Several research has suggested that treated wastewater-based microalgae biomass be incorporated into animal feed. Due to public opinion and quality food rules on animal feeds, it hasn't drawn much notice.

6.5.3 FOR BIOFUEL PRODUCTION

Biofuel is a material (biohydrogen, biodiesel, bioethanol, and biomethanol) with a high heat of combustion that is derived from biomass. The need for energy is rising on a global scale. Around 80% of the energy we use comes from fossil fuels. However, burning and extracting fossil fuels raises the number of greenhouse gases released into the atmosphere, which causes global warming. Due to the problems with fossil fuels, studies have recently concentrated on alternate energy sources. As an alternative energy source, biofuels have received a lot of attention and are seen as having a lot of potential.

It has been suggested that microalgae present the most significant potential for use in the manufacture of biofuels. This is due to their rapid growth and great photosynthetic efficiency. First-generation biofuels prompted concerns about a lack of arable land, which may lead to global food scarcity. Recently, interest in oil plants including palm, soybean, and rapeseed has increased, but their cultivation requires a lot of arable lands. *Jatropha curcas* is a plant used in second-generation biofuels, however, due to its slow growth rate and high demand for arable land, it is not economically viable. However, microalgae biofuel yields 10–100 times more gasoline per unit area than conventional crops like soya bean and oil palm. Because they may be cultivated in effluent from breweries and other industrial settings, microalgae do not cause any problems that are associated with the use of arable land (Deora et al., 2022).

6.6 ISSUES ASSOCIATED WITH MICROALGAL WASTEWATER TREATMENT

6.6.1 WASTEWATER PRETREATMENT

Microalgae growth is stunted by the abundance of bacteria, protozoa, fungi, and solid particles in raw brewery wastewater. In wastewater, these organisms compete with microalgae for nutrients and other minerals. Before the incorporation of microalgae, the wastewater must be pre-treated to remove any organisms. Various pretreatment technologies have recently been used to regulate large volumes of brewery wastewater (Gunes et al., 2019). Autoclaving and filtration are common pretreatment methods (Ramsundar et al., 2017). However, research has shown that autoclaving is the most efficacious pretreatment method for the removal of microorganisms. On the other hand, the authors also reported that autoclaving may disrupt the nutrient content of the wastewater. Cho et al. (2011) researched to validate this theory. When microalgae were grown in municipal effluent, the authors found that the filtration method produced more biomass than the autoclaving method. This indicates that the microalgae may not have been able to fully utilize light for photosynthetic activities due to the particulate matter present in the autoclaved pre-treated effluent. These pretreatment methods may not be commercially viable due to their high energy costs. Other alternative wastewater pretreatment methods, however, have been reported. For example, ultraviolet (UV) and chlorination have been identified as effective wastewater pretreatment methods by Qin et al. (2014).

6.6.2 SELECTION OF APPROPRIATE MICROALGAL STRAIN FOR BREWERY WASTEWATER TREATMENT

It is extremely important to select the appropriate species of microalgae for the treatment of wastewater in the brewery and elsewhere. Microalgae species should be resilient to environmental changes due to brewery wastewater's physical and chemical composition. Additionally, the species must be able to share metabolites to adapt to stress, withstand any attack by undesirable species, and overcome nutrient limitations. Only a few species of microalgae can survive in brewery wastewater treatment.

6.6.3 MICROALGAL HARVESTING TECHNIQUES FROM WASTEWATER

The separation of microalgae from effluent is one of the challenges associated with the use of microalgae in brewery wastewater treatment as well as in other types of industrial wastewater treatment (Gonçalves et al., 2017). This process requires energy. Flotation, centrifugation, flocculation, gravity sedimentation, filtration, and ultrasonication are all ways to harvest. The harvesting of microalgae is made more difficult by several factors, including the rapid growth rate of the algae, the small portion of algae present in the total suspension, the tiny size of a single cell, and the negative cell surface charge, all of which prevent the cells from forming larger and more easily harvestable particles. These combinations influence conventional separation techniques such as sedimentation, filtration, and microstraining, thereby increasing the price of algal biomass harvesting. However, more research is required to identify more straightforward, cost-effective, and efficient ways to harvest microalgal biomass from wastewater.

6.6.4 INTERNAL SHADING

Internal shadowing inhibits microalgae photosynthetic activity (Kang et al., 2019). Microalgae can rapidly reproduce within a day (24 hours) in effluent from breweries because this effluent contains a high concentration of nutrients; nevertheless, the duration of their reproduction within the log phase may be limited (3.5 hours) (Amenorfenyo et al., 2019). The dense upper culture limits light penetration into the water, so the effluent's rapid cell growth may reduce light access. The use of photobioreactors or raceway ponds could be a viable option for dealing with this matter. In raceway ponds, the microalgae at the bottom are moved close to the surface by the rotation of the shift paddle in the culture media. This allows the microalgae to absorb light energy. In a photobioreactor system, the light source is positioned so that it is close to the upper portion of the photobioreactor. At the same time, the air is sparged into the system, which rotates the lower portion of the microalgae so that it is closer to the surface. This allows the microalgae to absorb light energy.

6.7   SUMMARY AND CONCLUSIONS

Brewery industries consume significant amounts of water during the manufacturing process and discharge approximately 70% of it as effluent. The wastewater has high levels of COD, nitrogen, phosphorus, and other organic loadings, making it unfit for beneficial applications. Additionally, brewery wastewater has a high moisture content, high BOD concentration, and a temperature range of 25°C–38°C. The pH values vary depending on the concentration and category of chemicals used in cleaning and sanitizing wastewater. Several conventional treatment procedures have been employed to treat brewery wastewater. However, these conventional methods produce enormous volumes of sludge and have high costs for operation and maintenance, which makes them even less economical. Furthermore, excessive chemical use may result in ecological imbalances. The use of microalgae in wastewater treatment can help address environmental challenges and improve the overall quality of wastewater. Microalgae are a beneficial wastewater treatment agent due to their capacity to absorb nutrients and transform them into biomass. They can absorb nitrogen, phosphorus, and other nutrients in wastewater, releasing oxygen that bacteria use and fixing CO_2 through respiration. Microalgae prospects in wastewater treatment include cost-efficiency, minimal energy demand, abatement in sludge development, and emission of greenhouse gases. Brewery wastewater contains organic loads that are conducive to microalgae growth, making it an attractive option for environment-friendly and inexpensive wastewater treatment. Even though the microalgae approach may be able to remove a lot of impurities, the method has limitations when it comes to the removal of salt, odor, and color from BE. This technology must be used in conjunction with other affordable techniques to completely remove all toxins from the effluent. Membrane technology can be combined with microalgae-based technology for the polishing stage of wastewater treatment. Temperature and light conditions must be ideal for microalgae to grow. This technology might not function in temperate regions because of the low levels of sunshine and temperature. Artificial lighting systems could be used as an alternative, although doing so might increase the expense of healthcare in these countries. However, biomass can be used to create useful products like biodiesel. Thus, the utilization of microalgae in the treatment of brewery wastewater can enhance the economic viability and environmental sustainability of wastewater bioremediation.

The environmental damage caused by the brewing business is growing in tandem with the volume of wastewater it generates. Due to their efficacy in removing both high organic loads and COD, biological procedures like aerobic and anaerobic treatments are widely utilized for environmental protection. These methods are, however, connected with substantial capital and operating expenses. The AOP, activated carbon, and microbial fuel cell technologies have shown some encouraging findings in this research and offer a lot of potential for treating brewery wastewater.

High maintenance costs and energy usage, however, can operate as a deterrent. Membrane filtration is used to treat industrial wastewater from breweries and other industries. In addition, the technology is being utilized in the reuse of wastewater and drinking water. In recent years, this technique has seen rapid progress in terms of both quality and costs, and it might be utilized as a polishing step following the treatment with microalgae. Although activated carbon-based treatment systems are less expensive, effective at removing organic pollutants, and may be a good solution for the brewing industry, their extensive use of carbon/coal for wastewater treatment raises environmental and health risks. Microalgae treatment approaches have yielded encouraging results, according to this study. The treatment of brewery wastewater with microalgae has a significant amount of untapped potential. This tech is trustworthy, sustainable, and affordable. Additionally, it is excellent in eliminating ammonia as well as phosphorus from the effluents of breweries. The multiple advantages of using microalgae to treat wastewater have been discussed in this article. However, this technology necessitates the incorporation of additional treatment techniques to enhance the final effluent for environmental protection. There are currently few works on these integration options. In particular, the removal of color and odor from BE for potential reuse necessitates urgent research into additional treatment methods using microalgae. Further scientific research should concentrate on identifying other microalgae strains that are resilient enough to adapt to stress and other growth inhibitors to efficiently treat breweries and other industrial wastewaters for environmental protection.

REFERENCES

Abdelfattah, A., Ali, S. S., Ramadan, H., El-Aswar, E. I., Eltawab, R., Ho, S. H., ... & Sun, J. (2022). Microalgae-based wastewater treatment: Mechanisms, challenges, recent advances, and future prospects. *Environmental Science and Ecotechnology*, *13*, 100205.

Ahammad, S. Z., Graham, D. W., & Dolfing, J. (2020). Wastewater treatment: Biological. In *Managing Water Resources and Hydrological Systems* (pp. 561–576). CRC Press.

Amenorfenyo, D. K., Huang, X., Zhang, Y., Zeng, Q., Zhang, N., Ren, J., & Huang, Q. (2019). Microalgae brewery wastewater treatment: Potentials, benefits and the challenges. *International Journal of Environmental Research and Public Health*, *16*(11), 1910.

Bouchareb, R., Bilici, Z., & Dizge, N. (2021). Potato processing wastewater treatment using a combined process of chemical coagulation and membrane filtration. *CLEAN-Soil, Air, Water*, *49*(11), 2100017.

Cho, S., Luong, T. T., Lee, D., Oh, Y. K., & Lee, T. (2011). Reuse of effluent water from a municipal wastewater treatment plant in microalgae cultivation for biofuel production. *Bioresource Technology*, *102*(18), 8639–8645.

Deora, P. S., Verma, Y., Muhal, R. A., Goswami, C., & Singh, T. (2022). Biofuels: An alternative to conventional fuel and energy source. *Materials Today: Proceedings*, *48*, 1178–1184.

Gao, J. M., Wang, B., Li, W., Cui, L., Guo, Y., & Cheng, F. (2023). High-efficiency leaching of Al and Fe from fly ash for preparation of polymeric aluminum ferric chloride sulfate coagulant for wastewater treatment. *Separation and Purification Technology*, *306*, 122545.

Gonçalves, A. L., Pires, J. C., & Simões, M. (2017). A review on the use of microalgal consortia for wastewater treatment. *Algal Research*, *24*, 403–415.

Gunes, B., Stokes, J., Davis, P., Connolly, C., & Lawler, J. (2019). Pre-treatments to enhance biogas yield and quality from anaerobic digestion of whiskey distillery and brewery wastes: A review. *Renewable and Sustainable Energy Reviews*, *113*, 109281.

Han, X., Hu, X., Yin, Q., Li, S., & Song, C. (2021). Intensification of brewery wastewater purification integrated with CO_2 fixation via microalgae co-cultivation. *Journal of Environmental Chemical Engineering*, *9*(4), 105710.

Harirchi, S., Wainaina, S., Sar, T., Nojoumi, S. A., Parchami, M., Parchami, M., ... &Taherzadeh, M. J. (2022). Microbiological insights into anaerobic digestion for biogas, hydrogen or volatile fatty acids (VFAs): A review. *Bioengineered*, *13*(3), 6521–6557.

Haq, I., & Kalmdhad, A. S. (2023). Enhanced biodegradation of toxic pollutants from paper industry wastewater using *Pseudomonas* sp. immobilized in composite biocarriers and its toxicity evaluation. *Bioresource Technology Reports*, *24*, 101674.

Haq, I., Kalamdhad, A. S., & Pandey, A. (2022). Genotoxicity evaluation of paper industry wastewater prior and post-treatment with laccase producing *Pseudomonas putida* MTCC 7525. *Journal of Cleaner Production*, *342*, 130981.

Haq, I., & Kalamdhad, A. S. (2021). Phytotoxicity and cyto-genotoxicity evaluation of organic and inorganic pollutants containing petroleum refinery wastewater using plant bioassay. *Environmental Technology & Innovation*, *23*, 101651.

Haq, I., & Raj, A. (2018). Markandeya Biodegradation of Azure-B dye by *Serratia liquefaciens* and its validation by phytotoxicity, genotoxicity and cytotoxicity studies. *Chemosphere*, *196*, 58–68.

Haq, I., Kumar, S., Kumari, V., Singh, S. K., & Raj, A. (2016a). Evaluation of bioremediation potentiality of ligninolytic *Serratia liquefaciens* for detoxification of pulp and paper mill effluent. *Journal of Hazardous Materials*, *305*, 190–199.

Haq, I., Kumar, S., Raj, A., Lohani, M., & Satyanarayana, G. N. V. (2017). Genotoxicity assessment of pulp and paper mill effluent before and after bacterial degradation using *Allium cepa* test. *Chemosphere*, *169*, 642–650.

Haq, I., Kumari, V., Kumar, S., Raj, A., Lohani, M., & Bhargava, R. N. (2016b). Evaluation of the phytotoxic and genotoxic potential of pulp and paper mill effluent using *Vigna radiata* and *Allium cepa*. *Advanced Biology*, 2016, 1–10, Article ID 8065736.

He, Y., Lian, J., Wang, L., Su, H., Tan, L., Xu, Q., ... & Hu, Q. (2022). Enhanced brewery wastewater purification and microalgal production through algal-bacterial synergy. *Journal of Cleaner Production*, *376*, 134361.

Huang, H., Schwab, K., & Jacangelo, J. G. (2009). Pretreatment for low pressure membranes in water treatment: A review. *Environmental Science & Technology*, *43*(9), 3011–3019.

Jaiyeola, A. T., & Bwapwa, J. K. (2016). Treatment technology for brewery wastewater in a water-scarce country: A review. *South African Journal of Science*, *112*(3-4), 1–8.

Kabir, S. B., Khalekuzzaman, M., Hossain, N., Jamal, M., Alam, M. A., & Abomohra, A. E. F. (2022). Progress in biohythane production from microalgae-wastewater sludge co-digestion: An integrated biorefinery approach. *Biotechnology Advances*, *57*, 107933.

Kang, D., Kim, K. T., Heo, T. Y., Kwon, G., Lim, C., & Park, J. (2019). Inhibition of photosynthetic activity in wastewater-borne microalgal-bacterial consortia under various light conditions. *Sustainability*, *11*(10), 2951.

Khan, S., Naushad, M., Iqbal, J., Bathula, C., & Sharma, G. (2022). Production and harvesting of microalgae and an efficient operational approach to biofuel production for a sustainable environment. *Fuel*, *311*, 122543.

Kusmayadi, A., Leong, Y. K., Yen, H. W., Huang, C. Y., & Chang, J. S. (2021). Microalgae as sustainable food and feed sources for animals and humans-biotechnological and environmental aspects. *Chemosphere*, *271*, 129800.

Lewoyehu, M. (2021). Comprehensive review on synthesis and application of activated carbon from agricultural residues for the remediation of venomous pollutants in wastewater. *Journal of Analytical and Applied Pyrolysis*, *159*, 105279.

Li, J., Otero-Gonzalez, L., Michiels, J., Lens, P. N., Du Laing, G., & Ferrer, I. (2021). Production of selenium-enriched microalgae as potential feed supplement in high-rate algae ponds treating domestic wastewater. *Bioresource Technology*, *333*, 125239.

Li, K., Liu, Q., Fang, F., Luo, R., Lu, Q., Zhou, W., ... & Ruan, R. (2019). Microalgae-based wastewater treatment for nutrients recovery: A review. *Bioresource Technology, 291,* 121934.

Li, Y., Horsman, M., Wu, N., Lan, C. Q., & Dubois-Calero, N. (2008). Biofuels from microalgae. *Biotechnology Progress, 24*(4), 815–820.

Liu, J., Deng, S., Qiu, B., Shang, Y., Tian, J., Bashir, A., & Cheng, X. (2019). Comparison of pretreatment methods for phosphorus release from waste activated sludge. *Chemical Engineering Journal, 368,* 754–763.

Lutzu, G. A., Zhang, W., & Liu, T. (2016). Feasibility of using brewery wastewater for biodiesel production and nutrient removal by Scenedesmus dimorphus. *Environmental Technology, 37*(12), 1568–1581.

Mata, T. M., Melo, A. C., Simões, M., & Caetano, N. S. (2012). Parametric study of a brewery effluent treatment by microalgae Scenedesmus obliquus. *Bioresource Technology, 107,* 151–158.

Menkiti, M. C., Sekeran, G., Ugonabo, V. I., Menkiti, N. U., &Onukwuli, O. D. (2016). Factorial optimization and kinetic studies of coagulation-flocculation of brewery effluent by crab shell coagulant. *Journal of the Chinese Advanced Materials Society, 4*(1), 36–61.

Niu, D., Yuan, X., Cease, A. J., Wen, H., Zhang, C., Fu, H., & Elser, J. J. (2018). The impact of nitrogen enrichment on grassland ecosystem stability depends on nitrogen addition level. *Science of the Total Environment, 618,* 1529–1538.

Okolo, B. I., Nnaji, P. C., & Onukwuli, O. D. (2016). Nephelometric approach to study coagulation-flocculation of brewery effluent medium using *Detarium microcarpum* seed powder by response surface methodology. *Journal of Environmental Chemical Engineering, 4*(1), 992–1001.

Polprasert, C. (2007). *Organic Waste Recycling: Technology and Management.* IWA Publishing, London.

Qin, L., Shu, Q., Wang, Z., Shang, C., Zhu, S., Xu, J., ... & Yuan, Z. (2014). Cultivation of Chlorella vulgaris in dairy wastewater pretreated by UV irradiation and sodium hypochlorite. *Applied Biochemistry and Biotechnology, 172,* 1121–1130.

Raji, M., Mirbagheri, S. A., Ye, F., & Dutta, J. (2021). Nano zero-valent iron on activated carbon cloth support as Fenton-like catalyst for efficient color and COD removal from melanoidin wastewater. *Chemosphere, 263,* 127945.

Ramsundar, P., Guldhe, A., Singh, P., & Bux, F. (2017). Assessment of municipal wastewaters at various stages of treatment process as potential growth media for *Chlorella sorokiniana* under different modes of cultivation. *Bioresource Technology, 227,* 82–92.

Ranjit, P., Jhansi, V., & Reddy, K. V. (2021). Conventional wastewater treatment processes. In *Advances in the Domain of Environmental Biotechnology: Microbiological Developments in Industries, Wastewater Treatment and Agriculture,* 455–479. 10.1007/978-981-15-8999-7_17.

Sawant, S. S., Gosavi, S. N., Khadamkar, H. P., Mathpati, C. S., Pandit, R., & Lali, A. M. (2019). Energy efficient design of high depth raceway pond using computational fluid dynamics. *Renewable Energy, 133,* 528–537.

Sekine, M., Mizuno, N., Fujiwara, M., Kodera, T., & Toda, T. (2022). Improving methane production from food waste by intermittent agitation: Effect of different agitation frequencies on solubilization, acidogenesis, and methanogenesis. *Biomass and Bioenergy, 164,* 106551.

Singh, R., Bhunia, P., & Dash, R. R. (2017). A mechanistic review on vermifiltration of wastewater: Design, operation and performance. *Journal of Environmental Management, 197,* 656–672.

Sinha, S., Nigam, S., & Syed, M. (2022). Insight into advanced oxidation processes for wastewater treatment. In *Advanced Oxidation Processes for Wastewater Treatment* (pp. 101–106). CRC Press.

Sirohi, R., Pandey, A. K., Ranganathan, P., Singh, S., Udayan, A., Awasthi, M. K., ... & Sim, S. J. (2022). Design and applications of photobioreactors-A review. *Bioresource Technology*, *349*, 126858.

Song, C., Hu, X., Liu, Z., Li, S., & Kitamura, Y. (2020). Combination of brewery wastewater purification and CO2 fixation with potential value-added ingredients production via different microalgae strains cultivation. *Journal of Cleaner Production*, *268*, 122332.

Song, Y., Wang, L., Qiang, X., Gu, W., Ma, Z., & Wang, G. (2022). The promising way to treat wastewater by microalgae: Approaches, mechanisms, applications and challenges. *Journal of Water Process Engineering*, *49*, 103012.

Subramaniyam, V., Subashchandrabose, S. R., Ganeshkumar, V., Thavamani, P., Chen, Z., Naidu, R., & Megharaj, M. (2016). Cultivation of Chlorella on brewery wastewater and nano-particle biosynthesis by its biomass. *Bioresource Technology*, *211*, 698–703.

Thakkar, A. P., Dhamankar, V. S., Kapadnis, B. P. (2006). Biocatalytic decolourisation of molasses by Phanerochaetechrysosporium. *Bioresource Technology*, 97(12), 1377–81.

Tonhato Junior, A., Hasan, S. D. M., & Sebastien, N. Y. (2019). Optimization of coagulation/flocculation treatment of brewery wastewater employing organic flocculant based of vegetable tannin. *Water, Air, & Soil Pollution*, *230*, 1–18.

Travieso, L., Benítez, F., Sánchez, E., Borja, R., León, M., Raposo, F., & Rincón, B. (2008). Assessment of a microalgae pond for post-treatment of the effluent from an anaerobic fixed bed reactor treating distillery wastewater. *Environmental Technology*, *29*(9), 985–992.

Wang, B., Li, Y., Wu, N., & Lan, C. Q. (2008). CO$_2$ bio-mitigation using microalgae. *Applied Microbiology and Biotechnology*, *79*, 707–718.

Zou, Y., Zeng, Q., Li, H., Liu, H., & Lu, Q. (2021). Emerging technologies of algae-based wastewater remediation for bio-fertilizer production: A promising pathway to sustainable agriculture. *Journal of Chemical Technology & Biotechnology*, *96*(3), 551–563.

7 Environmental Pollution from Industrial Wastewater and its Bioremediation

Dharani Loganathan,
Judith Infanta Madhalai Muthu,
and Gokila Muthukrishnan

7.1 INTRODUCTION

Every living thing needs water to function, and it is necessary at every stage of the life cycle to meet a variety of needs. Water is employed for washing, food preparation, industrial activities, and agricultural tasks in addition to drinking. Water cannot now be replaced everywhere on the surface of the globe. Even though there is a lot of water in the oceans, it is not appropriate for everyday consumption. Consequently, the scenario has created a significant difficulty for usable water due to population growth, industry expansion, and climate change. The main sources of contamination is the increase in pollution in rivers and other water bodies brought on by the discharge of wastewater from various businesses (Xiong et al. 2019; Figure 7.1).

One of the biggest environmental issues caused by humans is industrial wastewater pollution. It is brought on by contaminants that are discharged into bodies of water by commercial operations like production, processing, and cleaning (Li et al. 2009). Heavy metals, suspended particles, dangerous compounds, and other pollutants that can seriously harm the environment and people's health can be found in industrial effluent. Bioremediation is one strategy for dealing with this pollution problem. Toxins and other pollutants found in wastewater can be broken down utilizing microorganisms, either naturally occurring or created through genetic engineering. This procedure is referred to as bioremediation. Bioremediation is a safe and economical method of treating contaminated water because it makes use of the enzymes produced by these bacteria to lower the concentration of hazardous chemicals. Typically, the process of bioremediation starts with the addition of particular bacteria to the contaminated water. These bacteria can more quickly degrade contaminants because they are uniquely suited to break down specific kinds of organic and inorganic particles. For instance, species of fungal spores can be used to lower the amount of heavy metals in liquid samples, while bacteria like Pseudomonas putida can be utilized to break down oil and other petroleum hydrocarbons. Additionally, these organisms' enzymes

DOI: 10.1201/9781003368120-7

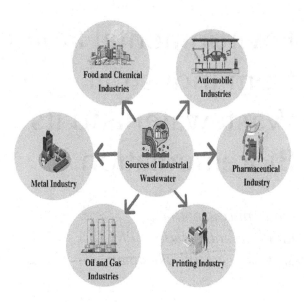

FIGURE 7.1   Sources of industrial wastewater.

can be employed to degrade other compounds including harmful poisons and chemicals that break down garbage. There are many benefits to the bioremediation process. It helps to lower the risk of health issues for humans and other animals by minimizing the amount of contaminants in wastewater. By preventing dangerous pollutants from getting into water bodies and aquatic ecosystems, it also helps to keep them safe for plant and animal life. Finally, it lessens the chance that wastewater may contaminate groundwater by filtering out hazardous substances before it is released back into the environment (Haq and Kalamdhad 2021, 2023; Haq et al. 2016a, b, 2017, 2022; Haq and Raj 2018). In general, bioremediation is a successful and effective method for lowering the concentrations of industrial contaminants in water. This procedure can aid in the breakdown of hazardous compounds and lessen their environmental impact by introducing certain bacteria and other microbes into contaminated sources (Inamuddin, Ahamed et al. 2020). Additionally, bioremediation can assist in lowering the cost associated with the treatment of wastewater. Bioremediation can help to save money while still offering a safe and efficient solution to remediate contaminated water by lowering the need for pricey chemical treatments. Ultimately, to address environmental deterioration brought on by industrial wastewater, bioremediation can be utilized alone or in addition with other traditional water treatment methods. In additional fields including waste management, oil spills, and agricultural runoff, bioremediation also offers a wide range of possible applications. The toxins can be eliminated and the water can be made safe again by introducing particular bacteria and other microbes into these sources of contamination. In addition, technologies for bioremediation are being developed to handle nuclear waste as well as soil and air

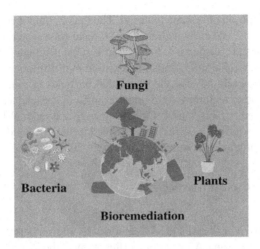

FIGURE 7.2   Principle of bioremediation.

pollutants. In the end, bioremediation is a critical tool for resolving many environmental issues brought on by human activity, and the possible applications are only constrained by our capacity to create new applications for it (Figure 7.2).

7.2   CHARACTERISTICS OF INDUSTRIAL WASTEWATER

7.2.1   PHYSICAL CHARACTERISTICS

7.2.1.1   Color

Domestic wastewater often ages according to its color. Fresh wastewater is typically gray, but septic wastes give the medium a black look. Numerous coloring agents may be found in industrial waste. Avoiding color in wastewater discharge from either home or industrial sources is crucial (Muttamara 1996).

7.2.1.2   Odor

The detection of odor has grown in significance as public concern about the efficient running of wastewater treatment plants has increased. Fresh wastewater typically has a pleasant smell, but when it decomposes biologically under anaerobic circumstances, a number of odorous chemicals are generated (Lovely et al. 1997).

7.2.1.3   Total Dissolved Solids

The total solids in wastewater is mainly made of both insoluble or suspended solids and soluble substances that are dissolved in water. By drying and weighing the residue left behind after screening the sample, it is possible to determine how much suspended particles are present in the effluent. Typically, the volatile solids are burned off when this residue is ignited. Although some organic material does not burn and certain inorganic salts decompose at high temperatures, volatile solids are typically

assumed to be composed of organic material. The major components of organic matter are lipids, proteins, and carbohydrates. In a typical wastewater, between 40% and
65% of the particles are suspended. Solids that can be eliminated by sedimentation
are referred to as settleable solids and are measured in milliliters per liter. According
to observations, typically 60% of the suspended particles in municipal wastewater
can be settled. Another technique to divide the solids in wastewater is into those that
volatilize at high temperatures (600°C) and those that do not. While the latter are
referred to as fixed solids, the earlier are identified as volatile solids. The majority of
volatile solids are organic (Chethana Krishna et al. 2014).

7.2.2 CHEMICAL CHARACTERISTICS

7.2.2.1   pH

The range of industrial effluent's pH value normally ranges between 4.44 and 8.90,
with a mean of 7.23. Sana Aluminum's effluents had the lowest pH, while Pepsi
Industries had the highest pH. When compared to NEQS guidelines, the pH value of
Sana Aluminum's effluents was over the allowable limit and may negatively impact
aquatic life because of its extremely acidic nature (Tariq et al. 2015).

7.2.2.2   Organic Materials

The main components of organic molecules are typically carbon, hydrogen, oxygen,
and, in certain situations, nitrogen. There may also be other crucial components including iron, phosphorus, and sulfur. Protein makes up 40%–60% of the organic compounds in wastewater, followed by carbohydrates, which make up 25%–50%, and fats
and oils, which make up 10%. Urine's main component, urea, is another significant
organic substance that contributes to wastewater. Under-composed urea is not typically
seen in wastewaters because of how quickly it decomposes. In addition, a vast variety
of synthetic element or compound structures can be found in wastewater in modest
amounts. The oxygen supplies of the receiving water will be diminished by the presence of easily biodegradable organic materials. The techniques used to treat wastewater are complicated by the presence of organic material that is either non-biodegradable
or difficult to decompose. This non-degradable fraction has a number of drawbacks,
including the possibility of harming aquatic species, the transferring of color and odor
to receiving water, and greater treatment costs required for downstream use.

7.2.2.3   Inorganic Materials

Common inorganic components found in typical wastewater include chloride, hydrogen ions, alkalinity-inducing substances, nitrogen, phosphorus, sulfur, and heavy
metals. Because of these chemicals' growth-restraining or eutrophic properties, even
minimal concentrations of them can have a significant impact on organisms in receiving waters. The ability to use inorganics as a substrate in their metabolism exists in
both algae and macrophytic plant types. Carbon, nitrogen, and phosphorus, the so
called main elements of inorganic metabolite seem to be present in natural streams in
forms that are useful to plant life. Phosphorus is frequently present in natural streams
at much lower amounts than nitrogen or carbon. However, phosphorus has to be present in very small concentrations in order to promote algal growth (Muttamara 1996).

7.3   AN OVERVIEW OF VARIOUS WASTEWATER TREATMENTS

7.3.1   WASTEWATER TREATMENT

There are several potential sources of water contamination, including residences, industry, mining, and penetration, but one of the biggest is still related to industry's extensive usage of water. The distinction between water is typically made between four categories: rainwater, home wastewater, agricultural water, and industrial wastewaters. The final category can be further broken down into cooling water, washing effluent (which might vary in composition), and manufacturing or process water (which may be poisonous or biodegradable). Process waters, often known as wastewaters or effluents, typically present the biggest issues. In one crucial sense, drinking water sources (often streams, lakes, or ponds) and wastewaters are very different from one another. In comparison to wastewater from industrial-type operations, the majority of sources of drinking water have relatively low pollutant levels. Since the effluent contains different sorts of pollutants based on where it came from, the issues that arise throughout wastewater treatment are generally highly complex. Therefore, there are various effluent kinds to address, and each has unique properties necessitating different treatment techniques (Crini et al. 2018).

7.3.2   INDUSTRIAL WASTEWATER TREATMENT—CURRENT PRACTICES

The ability to extract value-added raw resources from industrial wastewater has been increased with the advancement of science and technology. When biodegradable materials were treated, suspended particles were removed, and bacteria were eliminated, wastewater treatment got underway around 1900. And as time has gone on, techniques have been developed to safeguard the environment's aesthetic value and lessen their detrimental effects on human health. Many advances and breakthroughs in hybrid methodology—which combines two to three processes—have been made in the course of extensive study employing physical, chemical, and biological processes to achieve high recovery rates and increased efficiency (Dutta et al. 2021).

7.3.3   PHYSICAL TREATMENT METHODS

7.3.3.1   Preliminary Treatment

The aim of preliminary treatment is to get rid of coarse particles and other big debris that are frequently present in untreated wastewater. Large, entrained, suspended, or floating materials can be removed or reduced in size with the use of preliminary treatment. Along with some fecal matter, these solids include bits of wood, cloth, paper, plastic, rubbish, and other materials. Heavy inorganic materials like rock and gravel, as well as any plastic or glass, are removed.

7.3.3.2   Primary Treatment

Primary treatment uses the physical techniques of flotation and sedimentation to remove both organic and inorganic particles. During primary treatment, between 25% and 50% of the entering biochemical oxygen demand (BODs), between 50% and

70% of the total suspended solids (SS), and 65% of the oil and grease are removed. Primary sedimentation also removes some organic phosphorus, organic nitrogen, and heavy metals that are linked to solids, but it has no effect on colloidal or dissolved elements. Primary effluent is the wastewater from primary sedimentation units.

7.3.3.3 Secondary Treatment

The major objective of secondary treatment is the removal of all suspended and organic matter from the effluent from initial treatment. There are around 65% suspended solids, 30% dissolved solids, and 6% colloidal solids in the mixture. To remove as much suspended particles as is practical is the aim of first treatment. For first treatment, the organic and inorganic components in wastewater that can be settled out using settling tanks or clarifiers. Since they are colloidal and dissolved, the majority of the organic and inorganic components in the main treatment effluent are also. The level of organics removal from wastewater that is required by contemporary effluent regulations and water quality criteria cannot be achieved only by primary treatment. The secondary treatment procedure involves biologically treating wastewater in a controlled environment using a wide variety of microorganisms. For secondary treatment, a variety of aerobic biological processes are employed. These procedures vary principally in how oxygen is provided to the microorganisms and how quickly organisms digest organic materials (Sonune et al. 2004).

7.3.4 CHEMICAL TREATMENT METHODS

Chemical treatment techniques are another option to handle industrial wastewater pollution. These techniques entail the use of chemicals to remove contaminants from water. Examples include the emulsification of grease and oil removal or the filtering of suspended particulates. These techniques can be efficient in removing specific kinds of pollutants, but they can also be costly and generate a lot of chemical waste. Additionally, because chemical treatments might bring more pollutants into water bodies, they can harm aquatic ecosystems. In order to get the greatest environmental outcomes, chemical treatment techniques should normally be employed as a last resort and in conjunction with other techniques like bioremediation. This is true even if they can be beneficial in some situations (Nasr et al. 2004). To further increase the effectiveness and efficiency of wastewater treatment, chemical treatments can be combined with additional techniques like activated sludge or artificial wetlands. In order to degrade contaminants and remove them from water, activated sludge entails adding microorganisms to wastewater. Utilizing carefully created systems, constructed wetlands remediate tainted water by utilizing organic processes like evaporation and filtration. It is conceivable to develop a complete, cost-effective system for treating industrial wastewater contamination by combining these techniques with chemical treatment.

7.3.5 BIOLOGICAL TREATMENT METHODS

There are essentially two distinct biological process types used to treat water and wastewater. In the first of these, also known as suspended growth methods, the microorganisms in charge of the treatment process are maintained in liquid suspension by adequate mixing techniques. On the other hand, in attached growth processes, the microbes in responsible of digesting biological matter or nutrients were bound to a benign carrier system like rock, gravel, or dross, or to a diverse range of synthetic materials. Additionally, both techniques may be utilized as a component of an unitary treatment method (Chaudhary et al. 2019).

7.4 PRINCIPLE OF BIOREMEDIATION

7.4.1 BIOREMEDIATION

The process of bioremediation involves converting complicated contaminants into safe or nontoxic compounds, which can then be utilized as nutrients by other plants and microorganisms. Bioremediation is the process of cleaning up contaminated places. Bioremediation depends heavily on microorganisms like fungi, bacteria, algae, etc. Some microbes can't break down contaminants into basic, harmless molecules that can be consumed. Decontaminating the site is the primary objective of bioremediation. The most straightforward form of bioremediation is the addition of chemicals, nutrients, and oxygen to the water to aid the local microorganism in destroying the harmful substance in wastewater (Figure 7.3).

7.4.2 PRINCIPLE

The idea of biological remediation is built on biodegradation. Environmental waste can be organically decomposed in a safe state or to concentrations below the proper concentration limits using the bioremediation method. This is carried out under stringent guidelines established by the relevant regulatory organizations. Bioremediation transforms hazardous waste or chemicals into less dangerous or

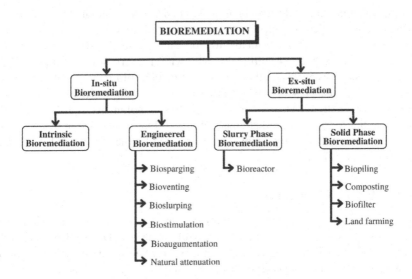

FIGURE 7.3 Classification of bioremediation.

nontoxic forms that can be found in wastewater by using naturally occurring living organisms, most often bacteria. Although it is a green treatment option and a commercially viable technology, its effectiveness may be impacted by the environment. Bacteria, fungi, or plants with the physiological capacity to break down or detoxify hazardous contaminants in water and purify it are the required microorganisms. It uses low-cost, onsite technologies. The development of supportive microflora or microbial consortia is a prerequisite for this method. They are accustomed to living in hazardous environments, so they can carry out important jobs. These microbial consortia can be produced in a variety of methods, such as by promoting growth, adding nutrients, including a terminal electron acceptor, or regulating the temperature and humidity. Microorganisms use these pollutants as food or energy sources during bioremediation processes. Under some conditions, certain natural bacteria may already be at the site, while others may be collected and added to the treated material via the bioreactor. It is important to realize that while bioremediation depends on the growth and activity of bacteria, its success also depends on the environmental factors that influence the growth and pace of microbial degradation. In order to find the appropriate microorganisms under the required environmental conditions, the bioremediation process often depends on doing so. It is feasible to strengthen the biological processes used in bioremediation technologies and concentrate on getting rid of dangerous compounds from water. This process turns trash into non-hazardous substances like water, carbon dioxide, and biomass, causing waste to mineralize and reducing the need for additional treatment (Ojha et al. 2021).

7.5 MICROBIAL POPULATION FOR BIOREMEDIATION PROCESS

7.5.1 Microbial Ecology

The removal of pollutants through bioremediation is a continuation of regular microbial metabolism. Microbial enzymes can only speed up viable conversions when acting as biocatalysts; they are unable to work in a thermodynamic deficit without the proper energy support. Additionally, only within a certain range of physiological conditions can bacteria and their enzymes function. Bio-remedial efforts cannot be successful in the presence of conditions that limit life or deactivate enzymes. Therefore, while the physico-chemical properties of pollutants and the metabolic capacity of microorganisms dictate the viability of biotransformation reactions, the actual biotransformation process also is determined by the current ambient conditions (Dott et al. 1989).

7.5.2 Microbial Population

Microorganisms can be extracted from almost any type of environmental situation. Microbes will adapt and flourish in a variety of conditions. Due to their versatility, microbes and other biological systems can be used to eliminate or reduce environmental risks. The following categories can be used to categorize these bacteria.

7.5.2.1 Aerobic

In the presence of oxygen, Pseudomonas and Mycobacterium are some examples of aerobic bacteria known for their ability to degrade. Pesticides and hydrocarbons, including alkanes and polyaromatic chemicals, have been observed to be degraded by these bacteria often. As their only source of carbon and energy, many of these bacteria rely exclusively on the contamination.

7.5.2.2 Anaerobic

Aerobic microorganisms are utilized more frequently than anaerobic bacteria when there is no oxygen present. The utilization of anaerobic bacteria to bioremediate PCBs, trichloroethylene (TCE), and chloroform in sediment samples is the subject of increasing amounts of research.

7.5.2.3 Ligninolytic Fungi

Fungi can break down a very wide variety of harmful or persistent environmental contaminants. Straw, sawdust, or corn cobs are examples of typical substrates.

7.5.2.4 Methylotrophs

Aerobic bacteria use methane to create carbon and energy. The enzyme that exists in the aerobic degradation pathway, methane monooxygenase, has a broad range of substrates and is potent against a number of chemicals, that also includes the chlorinated aliphatics like trichloroethylene.

7.5.3 FACTORS LIMITING MICROBIAL POPULATION

7.5.3.1 Technical Factors

7.5.3.1.1 Bioavailability

The pace at which microbial cells may break down toxins through bioremediation is determined by the speed in which contaminants are taken in, broken down, and transferred to the cell. Higher microbial exchange rates do not result in greater biotransformation rates as mass transfer is indeed a limiting condition. The majority of polluted soils and sediments appear to have this characteristic (Vidali 2001). For instance, even after 50 years, the soil-contaminating explosives did not go through the degradation process. Treatments that involved abrasive soil mixing and splitting up of the bigger soil particles greatly accelerated biodegradation. A variety of physical and chemical mechanisms, including sorption and desorption, diffusion, and dissolution, regulate how bioavailable a pollutant is. Because of the sluggish molecular diffusion to the degrading microorganisms, pollutants in soil are less bioavailable. At zero mass transfer rate, contaminants are no longer present. A common term for aging or weathering is the decline in bioavailability over time. It could be caused by a combination of things, including: (i) chemical oxidation reactions that incorporate contaminants into organic matter; (ii) slow diffusion into very microscopic gaps and absorption in and out of organic matter; and (iii) the development of semi-rigid non-aqueous-phase fluids (NAPL) with such a high resistance to NAPL-water molecular diffusion. The use of food-grade surfactants, that enhance the availability of pollutants for microbial degradation, can solve these bioavailability issues.

7.5.3.2 Non-technical Factors

A few of the non-technical requirements that affect bioremediation include the ability to meet the necessary clean-up goal, satisfactory cost in comparison to some alternative processes, acceptable threats in residual pollutants left after bioremediation, beneficial public perception, supportive regulatory perception, capacity to accomplish time constraints, and ability to conform to space constraints (Boopathy 2000).

7.6 BIOREMEDIATION STRATEGIES—IN SITU AND EX SITU BIOREMEDIATION

7.6.1 BIOREMEDIATION STRATEGIES

Bioremediation is the technique by which organic wastes are biologically reduced under controlled circumstances to a benign state or levels below concentration limits set by regulatory agencies. Bioremediation is, by definition, the process of converting environmental pollutants into less hazardous forms using living organisms, typically bacteria. To decompose or detoxify pollutants harmful to human health or the environment, it employs naturally occurring bacteria, fungi, or plants. The bacteria may be found naturally in a polluted region or they may have been isolated and transported to the polluted area. Living things undergo reactions as elements of their physiological processes that transform contaminant chemicals (Tripathi et al. 2018). In situ bioremediation and ex situ bioremediation are two subcategories that can be distinguished (Table 7.1).

7.6.2 IN SITU BIOREMEDIATION AND ITS TECHNIQUES

The biological treatment process is used to remove harmful chemicals and other substances from wastewater. We need to combine a variety of scientific and technical disciplines in order to provide the proper circumstances for microbial development and transform organic stuff into less hazardous materials (Paul et al. 2021). The "in situ" bioremediation techniques are chosen for cleaning up contaminated water because they use microbial systems' metabolic ability to remove target contaminants

---

TABLE 7.1
Bioremediation Techniques

Technique	Examples	Benefits
In Situ Bio-remediation	Intrinsic Bioremediation, Biosparging, Bioattenuation	Most cost efficient, Non-invasive, Relatively passive, Naturally attenuated process, Treats soil and water
Ex situ Bio-remediation	Biopiling, Land Farming, Windrow	Simple, Cost-efficient, self -heating, Rapid reaction rate, Can be done onsite.

from the environment without having to dig up contaminated samples beforehand. The primary goal of "in situ" remediation is to use a variety of treatment methods to return the damaged environment to its pre-contamination state. Using "in situ" bioremediation procedures, pollutants are treated where they can be biologically destroyed in their natural environment (Azubuike et al. 2016). These methods entail handling contaminated materials right where the contamination occurred. Since there is no need for excavation, the soil structure is not significantly disturbed. In theory, these approaches should be less pricey than ex situ bioremediation procedures because excavation operations are not necessary. Bioventing, biosparging, and phytoremediation are some examples of in situ bioremediation procedures that may be improved, whereas other approaches may be used without any improvement at all (intrinsic bioremediation or natural attenuation). Chlorine-based solvents, pigments, toxic substances, and hydrocarbon-polluted areas have all been effectively treated with in situ bioremediation approaches. Importantly, the availability of nutrients, moisture content, pH, and temperature are among the critical environmental factors that must be adequate for an in situ bioremediation to be performed.

7.6.2.1 Intrinsic Bioremediation

Natural attenuation, another name for intrinsic bioremediation, is an in situ bioremediation process that includes passive remediation of contaminated places without the need for any outside force (human intervention). Polluting chemicals, particularly those that are refractory, can be biodegraded via both aerobic and anaerobic microbial activities (Harrington 1978). Since it doesn't call for any intervention, this method is the preferred one for biotherapy. The approach enables the ecosystem to return to its pre-damaged state, prevents habitat damage, and facilitates the detoxification of harmful substances (Azubuike et al. 2016). Due to the lack of an external force, the procedure may be less expensive than other in situ techniques. To prove that bioremediation is continuing and sustained, however, the process must be observed; therefore, the phrase "monitored natural attenuation" (MNA). Due to the cold temperature conditions that are likely to have a detrimental impact on the biodegradation process, MNA is increasingly becoming accepted in the majority of European nations, with the exception of a very small number. Additionally, it has been suggested that the primary mechanism for the removal of pollutants via intrinsic bioremediation is biodegradation (MNA). The removal of polyaromatic hydrocarbons (PAHs), which would lessen the ecotoxicity of polluted soil, was also demonstrated as being insufficiently accomplished by intrinsic bioremediation.

7.6.2.2 Bioattenuation

The method depends on converting contaminants into harmless forms or an immobilized condition. According to physical and chemical principles, proteobacteria-mediated bioattenuation has been reported to be successful in removing metals like As, Ni, and Al, while acinetobacteria have been found to participate in the bioattenuation process for naturally removing metals.

7.6.2.3 Bioaugmentation

Enhanced consortia or strains are introduced during the bioaugmentation process when the necessary microbial cells are insufficient to degrade the target chemicals.

Deployment of a variety of naturally occurring microbial strains or a genetically modified variation to clean polluted soil or water When treating municipal wastewater in a bioreactor, it is used to restart the activated sludge process. The majority of cultures on the market are composed of a combination of microbial cultures developed via research and include all required microorganisms (Paul et al. 2021). According to reports, a number of biotic and abiotic factors are responsible for effective bioaugmentation, including the removal of heavy metals by a group of filamentous fungi using bioaugmentation practices and the production of cadmium-enriched water by Streptomyces species. To ensure that the in situ microorganisms can completely break down the chlorinated ethenes, such as tetra- and trichloro-ethylene, to the nontoxic compounds ethylene and chloride, bioaugmentation is utilized in places where water is polluted. While autochthonous microorganisms can change their environment by releasing organic compounds due to a variety of unfavorable biotic and abiotic factors, the progress of bioaugmentation can be increased by choosing the right operating strategies that are intended to increase the resilience and protracted effectiveness of inoculated microbes. It is infact challenging to monitor this system.

7.6.2.4 Bioslurping

Bioslurping is the concurrent use of vacuum-enhanced recovery, soil vapor extraction, and bioventing to promote land and water clean-up by obliquely integrating oxygen and promoting pollutant biodegradation. Light non-aqueous phase liquids (LNAPLs) are one type of free product recovery for which it is intended, without the need to remove significant amounts of wastewater. The method utilizes a "slurp" tube that reaches through into the free product layer and sucks up liquid from the layer in a style analogous to how a straw draws liquid up through any container. LNAPLs are propelled upward to the surface by the pumping mechanism, where they are cut off from water and air.

7.6.2.5 Bioventing

The main objective of bioventing is to use moisture and nutrient addition to help bioremediation by causing pollutants to undergo microbial transformation into a nontoxic state. Maintaining aerobic conditions and obtaining reasonable biodegradation rates are the two fundamental requirements for successful bioventing, and to meet these requirements, naturally occurring hydrocarbon-degrading microorganisms must be present in sufficient concentrations. As a result, air injection rate is one of the key parameters for pollutant dispersion. This method uses regulated airflow stimulation to promote bioremediation by boosting the activity of local bacteria by supplying oxygen to the unsaturated (vadose) zone. The ultimate objective of bioventing is to accomplish microbial transformation of contaminants into a harmless condition. Amendments are made by adding nutrients and moisture to increase bioremediation. In comparison to other in situ bioremediation methods, this one has gained favor, particularly for cleaning up oil spill sites. It was found that the increased airflow rates do not improve the pace of biodegradation or raise the efficiency of pollutant biotransformation.

This is a result of early air saturation in the subsurface due to biodegradation's requirement for oxygen. However, the modest air input rate significantly increased biodegradation. Thus, it demonstrates that air injection rate is one of the crucial variables affecting bioventing. The efficiency of bioventing-based bioremediation relies on air injection locations. Due to other environmental factors and unique properties, achieving results similar to those seen in laboratory studies during onsite field trials is not always possible. Therefore, bioventing may lead to prolonged treatment times.

7.6.2.6 Biosparging

Biosparging is the process of speeding up the biodegradation of pollutants by naturally existing bacteria. By forcing air under pressure below the water table, it oxidizes the aquifer, boosting the reaction rate and levels of oxygen. This method is very comparable to bioventing in that wastewater is given an air injection to encourage microbial activity and aid in pollutant removal from polluted environments. Contrary to bioventing, air is injected at the zone of saturation in air injection, which can induce an upward movement of volatile organic substances to the unsaturated zone to stimulate biodegradation. Pollutant biodegradability affects how well a biosparge works. In situ air sparging (IAS) uses high airflow rates to accomplish pollutant volatilization, but biosparging encourages biodegradation. Parallel to this, neither of the pollutant removal mechanisms for each technique is mutually exclusionary. Diesel and kerosene contamination of aquifers has been successfully treated with biosparging in many cases (Harrington 1978). When under stress, bacteria create compounds that absorb metals, which interact chemically with contaminants to precipitate them. Because oxygen input creates an aerobic environment conducive to local microbes' degradative action, these contaminants are broken down during biosparging. Significant flexibility in system design and installation is made possible by the tiny diameter air injection stations' simplicity and inexpensive setup costs (Table 7.2).

7.6.2.7 Biostimulation

In order to increase the microbial turnover of chemical pollutants, biostimulation of heavy metals depends on the presence of nutrients, oxygen, temperature, pH, and redox potential. Another method of biostimulation involves the insertion of

---

TABLE 7.2
Some Contaminant Sources of Water and their Bioremediation Process

Class of Contaminants	Example	Source	Process of Bio-remediation
Chlorinated solvents	Trichloroethylene	Drycleaners	Anaerobic
Polychlorinated biphenyls	4-Chlorobiphenyl	Electrical manufacturing	Anaerobic
Pesticides	Atrazine, 24D, Parathon	Agriculture, Pesticide manufacture	Both Aerobic and Anaerobic
BTEX	Benzene, Toluene, Ethyl benzene and Xylene	Oil production and storage, Gas works ites Airports, Paint manufacture	Both Aerobic and Anaerobic
Polyaromatic hydrocarbons	Naphthalene, Anthracene, Pyrene	Oil production and Storage, Engine works	Aerobic

biodegradable substances that act as major substrates, where the pollutant is removed as a subsidiary substrate but at controllable rates. Temperature control is only possible in designed systems since a warm environment promotes biomass activity, which is beneficial for biostimulation. Examples of these systems include enclosed vessel composting and slurry bioreactors.

7.6.3 Ex situ Bioremediation and its Techniques

This method of bioremediation involves treating contaminants away from the affected site. Composting is the process by which organic materials are broken down by microorganisms at extremely high temperatures. The typical compost temperature ranges from 55°C to 65°C. As the temperature rises, organic material decomposition into garbage accelerates. The following fundamental steps have been used to demonstrate windrow composting. To remove big rocks and other debris, polluted soils must first be dug and screened. The process of intervention known as "ex situ" bioremediation results in the breakdown of chemical contaminants that are found in the unearthed samples. In comparison to "in situ" techniques, "ex situ" remediation methods appear to be more expensive. These bioremediation systems exhibit notable variations in their experimental controls and the consistency of the process result. Since "ex situ" bioremediation is carried out in a non-natural setting, it can be manipulated with the aid of various physicochemical treatments. These methods entail removing pollutants from contaminated areas and then moving them to another location for treatment. The treatment cost, the depth of polluted air, the category of pollutant, the extent of pollution, the location, and the geology of the polluted site are typically taken into account when evaluating ex situ bioremediation techniques. A wider range of poisons is treated using "ex situ" bioremediation procedures, which are also simpler to control.

7.6.3.1 Biopiling

Nutrient enrichment and aeration are used in biopile-mediated bioremediation to promote microbial activity and hence enhance bioremediation. This specific "ex situ" technique is more frequently used due to its beneficial properties because nutrients, temperature, moisture, and pH are effectively managed to accelerate biodegradation. Because it is more effective than composting and land farming due to the mass transfer of nutrients and air, biopile is regarded as a better strategy for pollutant removal. Heavy metals from wastewater are removed using this method. A treatment bed, an aeration system, an irrigation/nutrient system, and a leachate collection system make up the fundamental biopile system. The irrigation/nutrient system is covered by applying positive pressure or suction from passing air and nutrients. To prevent evaporation, runoff, and volatilization and to promote solar heating, it can be up to 20 feet tall and covered with plastic (Paul et al. 2021). This technique includes a treatment bed, irrigation, nutrient and leachate collection systems, and aeration.

7.6.3.2 Bioreactors

The term "bioreactor" refers to a container created to use microbes to remove pollutants from wastewater. By using reactors and an engineered contaminated system, contaminated soil, sediment, or other materials can be bio-remediated. Slurry bioreactors are a type of containment vessel and equipment used to create a three-phase environment, or a combination of solid, liquid, and gas, in order to speed up the bioremediation of pollutants that are bound to soil and those that are water-soluble as a water slurry of biomass and contaminated soil and to remove the specific contaminants from it. The bioremediation rate is higher than in situ or solid-phase systems because the physical environment is more easily controlled and manageable (Dhanya 2021). The bioreactor can operate in batch, fed-batch, sequencing batch, multistage, or continuous modes. The overall market economy and capital expenditure determine the bioreactor's operating mode. The bioreactor maintains the environment of the cells to support their natural processes for supplying the best growth conditions. The parameters that are crucial for a bioreactor are temperature, pH, moisture, inoculum percentage, gas velocity, agitation rate, and substrate concentration (Azubuike et al. 2016). Although different operating parameters that can be easily controlled have made bioreactor-based bioremediation effective, determining the best operating conditions by relating all parameters using the one-factor-at-a-time (OFAT) approach would probably necessitate numerous experiments and take a lot of time. The bioremediation process in a bioreactor is constrained by mass transfer, controlled bioaugmentation, increased bioavailability of pollutants, and nutrient addition. A saturation transient that starts at the bottom of the column and moves upward characterizes the heavy metal removal process in an Anaerobic Upflow (ANFLOW) bioreactor. It was discovered that the fixed bed bioreactors may boost the effectiveness toward the removal of metals (Cu^{2+}, Pb^{2+}, Cd^{2+}, CO^{2+}, Ni^{2+}, Zn^{2+}, Mn^{2+}, and Fe^{3+}) in contaminated environmental samples by a strain of Pseudomonas aeruginosa to regulate various dangerous health hazard compounds. In a fixed packed-bed bioreactor with an alkaline pH, about 90% of the heavy metals were removed through biosorption. Due to a few factors, bioreactor-based bioremediation is not a specific method for removing heavy metals. First off, treating the vast volume of polluted soils or other substances and transporting pollutants to the place of treatment demand additional labor, money, and safety precautions, rendering this procedure inefficient and expensive. Second, because a bioreactor has so many different bioprocess variables if any of them is not properly controlled, it becomes a limiting factor, which lowers microbial activity and renders the technique ineffective. Thirdly, because different bioreactors react to pollutants in different ways, finding the best design is crucial. Most importantly, the expense of a bioreactor suitable for laboratory or pilot-scale bioremediation renders this method capitally expensive.

7.6.3.3 Windrow

By encouraging the activity of native and/or transitory hydrocarbonoclastic bacteria prevalent in contaminated water, and windrows, one of the ex situ bioremediation techniques, increases bioremediation. From prior Los Angeles open windrow composting experience, a windrow procedure for composting digested sludge was developed. In order to compost either digested or undigested sludge, a forced aeration method has been created. Multiple activities make up the windrow process, which is carried out in the open on a stable pad. In clearing and grubbing operations, woodchips from shredders are used as a bulking agent at a ratio of 1:3 by volume. Using a mechanical compost machine, windrows of woodchips and sludge are blended every day for around 2 weeks (Udaiyappan et al. 2017). The compost is piled up in a storage area after spreading and air drying to further stabilize it and get rid of pathogens. The material is filtered to eliminate the woodchips so they can be used again as a bulking agent after about 30 days in storage. After screening, the compost is prepared for use. The aerated pile process was created because initiatives to windrow-compost undigested sludge resulted in a high level of offensive odors, even though the windrow process is essentially odor-free when digested sludge is used. In this method, perforated metal pipes are installed on the ground and coated with woodchips or sludge that has previously been composted to a depth of 30 cm in order to absorb liquids and prevent the perforated pipes from becoming clogged. Over the chip-protected pipe, a pile of woodchips and sludge is built to a depth that is determined by the front-end loader that built the pile (Rigueto et al. 2020). The ratio of the mixture is the same as in the windrow method. Compost is spread over the entire pile in a layer that is 30 cm thick to insulate it and reduce odor leakage. For proper oxygen concentrations of between 5% and 15%, the vent pipe is linked to the centrifugal blower's suction side. The suction is provided at different rates. For odor control, air is taken out of the pile and fed through previously processed compost. Throughout the entire 21-day compost period, suction is applied sporadically in roughly 10-minute intervals. Condensate is said to flow off the pile at a rate of about 1 gal/ton (4 mL/kg) each day during the off cycle when the vacuum is lost. The compost is transferred to a stacked storage pile, just like with windrow composting, for curing and additional pathogen killing. After about 30 days of curing, the woodchips are screened to separate them. The composting facility essentially consists of an open air stabilized pad with drainage that diverts rainwater to a holding pond for treatment before being sprayed on nearby fields and woodlands Prospective plants can reduce runoff by offering an accessible, thatched structure to remove a significant percentage of rainwater from contact with the composting material, hence lowering the quantity and expense of polluted water treatment (Vishwakarma et al. 2020). Heavy metals like Cd and Ni have been discovered to be successfully removed by humic compounds. This technique may be used to fractionate metals including Cu, Zn, Fe, and Mn, as well as Cd. Windrow treatment demonstrated a higher rate of hydrocarbon removal when compared to biopile treatment.

7.6.3.4   Land Farming

Land farming is among the simplest bioremediation techniques due to its low cost and little equipment needs. It is recognized as an in situ bioremediation approach in some cases while being seen as ex situ bioremediation in others. The topic of contention is the treatment location. The depth of the contamination has a big impact on whether land farming can indeed be done in situ or ex situ. Dug or tilled unclean water is a common practice in land farming, however the sort of bioremediation seems to depend on the treatment site. It is feasible to treat excavated contaminated water in situ; otherwise, it must be treated ex situ because it is more similar to other ex situ bioremediation methods. According to reports, bioremediation can occur without excavation when a contaminant is less than 1 m below ground level, but it must be delivered to the ground surface for bioremediation to be boosted. The main practices that encourage the activity of native microorganisms to improve bioremediation during land farming are tillage, which results in aeration, addition of nutrients (nitrogen, phosphorous, and potassium), and irrigation. For the clean-up of hydrocarbon-polluted areas, especially those with polyaromatic hydrocarbons, land farming is typically employed. Therefore, the two remedial processes involved in pollution elimination are biodegradation and volatilization (weathering). The land farming method conforms to rules set out by the government and is adaptable to any environment. Leaching of pollutants into nearby regions is reduced during bioremediation operations by building a proper land agriculture design with an opaque liner. The land farming bioremediation technology, all things considered, is relatively easy to develop and implement, requires little capital investment, and can be utilized to remediate huge volumes of dirty soil with little negative environmental effect and energy demand. Despite being the simplest bioremediation method, landfarming has certain drawbacks: It necessitates a sizable operational area, more expense for excavation, a decrease in microbial activity because of adverse environmental circumstances, and less efficacy in inorganic pollution removal. Its design and pollutant removal (volatilization) method, particularly in hot (tropical) climatic zones, make it unsuitable for treating contaminated water containing harmful volatiles. Additionally, due to its construction and pollutant removal (volatilization) method, it is not appropriate for treating contaminated water containing harmful volatiles, especially in hot (tropical) climatic zones. Land farming-based bioremediation is time-consuming and far less productive than other ex situ bioremediation approaches due to these and other constraints. Ex situ bioremediation procedures have the significant benefit of not requiring a thorough evaluation of the contaminated site prior to treatment; this reduces the length, labor, and cost of the preparatory stage. Pollutant inhomogeneity resulting from depth, non-uniform concentration, and distribution may readily be reduced by successfully adjusting specific system parameters (temperature, pH, mixing), which are common to all ex situ techniques. The biological, chemical, and physicochemical conditions and parameters required for successful and efficient bioremediation can be changed using these strategies. Techniques for ex situ bioremediation are uncertain to be applied in particular locations, including as working sites, inner cities, and areas under buildings. Any ex situ bioremediation technology that requires moderate to significant engineering suggests that additional labor and resources are needed to build the technique. These procedures often need a lot of area to operate. Ex situ bioremediation methods may be utilized to address a variety of contaminants and are often quicker and simpler to regulate (Harrington 1978).

7.7   ADVANTAGES AND DISADVANTAGES OF BIOREMEDIATION

The main benefits and drawbacks of using bioremediation for wastewater treatment are addressed in Table 7.3.

7.8   CHALLENGES AND FUTURE PROSPECTS IN BIOREMEDIATION

Although it has become a viable alternative for water filtration, bioremediation still faces a number of obstacles that make it impractical for widespread commercial use (Nie et al. 2018). Because not all substances can be broken down by biological processes, bioremediation is severely constrained. If a substance is biodegradable at all,

TABLE 7.3

Benefits and Drawbacks of Bioremediation in Wastewater Treatment

Sl. NO.	Advantages	Disadvantages
1	Consumes less time than other techniques for treatment of polluted water	Applicable for compounds which are Biodegradable. All compounds cannot be degraded completely.
2	No transportation need	Difficult to extrapolate from bench and pilot-scale studies to full-scale field operations.
3	Less expensive technique	The products of biodegradation may be more persistent or toxic than the parent compound.
4	It uses natural microbes which is harmless to the environment compared to the use of hazardous chemicals.	Evaluating the performance of bioremediation is difficult and there are no acceptable endpoints for bioremediation treatments.

it may in some situations undergo additional processing and breakdown that results in the production of harmful material. Due to the occurrence of some site-specific limiting variables, a specific bacterial strain that is effective at one location may occasionally fail to operate well at other sites. The type of contaminants, the existence of an optimum quantity of nutrients, and the metabolic activities of microorganisms are thought to contribute to the intricacy of biological systems. However, using large machines and pumps may make some noise and cause some commotion. The procedure can occasionally be questioned in terms of its impact on other local microflora due to ethical concerns with the usage of certain bacterial strains (Ihsanullah et al. 2020). The majority of research described the use of bioremediation methods on a small scale to remove dyes. It is necessary to investigate the viability of bioremediation as an economical and effective technology for commercial use. Microbes' genomes might be modified to produce microbial strains with improved biodegradation properties. MFCs have become a viable alternative for bioremediation, although there is still work to be done to get beyond their drawbacks, such as their massive price and poor electricity output. There has been modest advancement in the production of bioenergy (such as hydrogen) from wastewater bioremediation. This innovative method not only lowers wastewater pollution but also stimulates the production of ecologically friendly gasoline. With the goal of potential commercialization, the synthesis of biohydrogen from wastewater utilizing microorganisms has to be further investigated. It is necessary to carefully assess the mechanisms of degradation, the operating circumstances, and the ideal microorganism growth environments. While the majority of research is based on synthetic wastewater, a small number of studies described the uses of bioremediation for the elimination of colors from actual wastewater. To determine their potential for use in real-world applications, it is crucial to investigate further the uses of bioremediation in wastewater. To properly address the existing challenges and broaden the commercial use of bioremediation methods for water purification, a multidisciplinary approach is necessary (Nie et al. 2018). Microalgae are tiny, single-celled organisms floating in water that have a hydrophilic (negative) surface charge. Gravitational settling is therefore challenging. Harvesting still has difficulties with low operational efficiency and high operating expenses as a result of its high energy efficiency as a step in the biodiesel manufacturing process. Even though some researchers have focused on industrial operations, research on microalgae removal of pesticides is now unable to get through the manufacturing cost barrier that is preventing commercial operations. In order to create products of high value, the majority of current studies are conducted at the medium or photobioreactor (PBR) level; nevertheless, the running expenses are very high (Divya et al. 2015). Under open operating circumstances, environmental elements like weather, temperature, light intensity, and external species naturally influence and limit the reproduction of microalgae. In order to increase the production of their own biomass, microalgae need the nutrients and contaminants in wastewater. Some particular nutrients can be used to increase economic output. However, certain harmful compounds that were ingested and accumulated cannot be completely destroyed and persist in the microalgal biomass. These compounds might influence subsequent operations and the downstream procedure if they are not correctly handled (Wu et al. 2014). Our efforts to comprehend and manage micro-societies with a cleaner, "green" concept have been significantly impacted by the developments made by recent molecular explorations in microbial technologies. It's likely that in the future, extremely efficient periphyton biofilms will be created by cloning the target microbe genome that removes contaminants (Maryjoseph et al. 2020). The quality of wastewater effluent following microalgae biomass separation is not well described in the literature. The identification and description of the side products produced throughout the bioremediation process must be the main focus of future study. Particularly, the relative toxicity of these byproducts and their existence may make it impossible to reuse the wastewater. It should be highlighted that the majority of the proof-of-concept investigations were carried out in batch reactors in a lab setting (Sutherland et al. 2019). In the real-world reactor system, it appears that the dynamics of operational factors including pH, temperature, dissolved oxygen, HRT, bacteria-microalgae interactions, light limitation/inhibition, and mixing conditions may greatly differ from the laboratory circumstances. Therefore, investigations at the pilot size are required to examine the difficulties in removing emerging contaminants (ECs) in continuous flow reactors in a dynamic environment (Sharma et al. 2021). A potential tactic is to maximize the usage of genetically modified microorganisms (GEM) to increase the capacity of bioremediation. This is because it is possible to combine new and effective metabolic pathways, expand the spectrum of substrates for existing pathways, and increase the stability of catabolic activity to build a designer biocatalyst target pollutant that includes resistant molecules. In an environmental application, an effective strategy involves concurrent gene transfer and GEM proliferation. Bacterial containment systems, which recreate any GEM that escapes from a contaminated environment. Additionally, using biological technology, a modified method of genetically modifying microorganisms with a specific polluted chemical might increase the effectiveness of bioremediation. By expanding the surface area and decreasing activation energy, nanomaterials lessen the toxicity of pollutants to microorganisms, which decreases the time and expense of bioremediation (Ojha et al. 2016).

7.9 SUMMARY AND CONCLUSION

Nowadays, there is a lot of worry about water contamination all around the world. Environmental degradation has been brought on by poor farming practices and fast industrialization as a result of increased human activity on energy supplies. Microorganisms are consequently essential for a significant alternative method of issue solving. A cutting-edge and ground-breaking method for cleaning up contaminated water is bioremediation. It is a healthy way to get rid of garbage. It is a very effective and popular tool for cleaning up contaminated environments. Thus, by better understanding microbial communities and how they interact with the environment and contaminants, as well as by learning more about the genetics of microorganisms, we can increase our ability to biodegrade pollutants. A variety of pollutants or harmful substances can be transformed into circumstances that are safe for humans or animals by the action of the right microorganisms. Microbes produce energy by breaking down contaminants. Eventually, the process might leave behind simpler molecules like carbon dioxide or water as well as cell biomass. This essentially eliminates any possible dangers associated with the handling and disposal of hazardous materials. Additionally, bioremediation may be done onsite, reducing the costs and dangers of shipping as well as any potential concerns to the environment and human health posed by the transportation of hazardous chemicals. The growing of microalgae in these wastewaters helps remediate high-nutrient-strength effluent by reusing the nutrients. Biofuel feedstock, fertilizers, and animal feed may all be made from the collected microalgae biomass. These wastewater treatment techniques struggle with internal shading, high suspended solid concentrations, and the harvesting and recovery of microalgae for use in downstream processes. PBRs or raceway ponds can be used to apply turbulent flow to solve these problems. Low energy consumption and low operating and maintenance costs are produced by the use of microalgae systems in the treatment of industrial wastewater. Due to the impact on the environment and public health, sewage effluent treatment is crucial. In order to maximize growing conditions, promote bio-adsorption, improve biodegrading enzymes, and use microalgal species for EC bioremediation qualities, more study is required. Future prospects include utilizing low-cost raw materials, optimizing medium, using molecular and genetic engineering to enhance strains, and modeling.

REFERENCES

C. C. Azubuike, et al (2016). "Bioremediation techniques-classification based on site of application: Principles, advantages, limitations and prospects", *World Journal of Microbiology and Biotechnology*, 32, 1–18.

R. Boopathy (2000). "Factors limiting bioremediation technologies", *Bioresource Technology* 74, 63–67.

D. K. Chaudhary, et al. (2019). "New insights into bioremediation strategies for oil-contaminated soil in cold environments", *International Biodeterioration & Biodegradation*, 142, 58–72.

P. Chethana Krishna (2014). "A research on cocoa pod husk activated carbon for textile industrial wastewater colour removal", *International Journal of Research in Engineering and Technology*, 3, 731–733.

G. Crini, et al. (2018). "Advantages and disadvantages of techniques used for wastewater treatment", *Environmental Chemistry Letters*, 17, 145–155.

M. S. Dhanya (2021). "Biosurfactant-enhanced bioremediation of petroleum hydrocarbons: Potential issues, challenges, and future prospects", *Bioremediation for Environmental Sustainability*, pp. 215–250. https://doi.org/10.1016/B978-0-12-820524-2.00010-9.

M. Divya, et al. (2015). "Bioremediation - An eco-friendly tool for effluent treatment: A Review", *International Journal of Applied Research*, 1(12), 530–537.

W. Dott, et al. (1989). "Comparison of autochthonous bacteria and commercially available cultures with respect to their effectiveness in fuel oil degradation", *Journal of Industrial Microbiology*, 4, 365–374.

D. Dutta, et al. (2021). "Industrial wastewater treatment: Current trends, bottlenecks, and best practices", *Chemosphere*, 285, 1–14.

W. M. Harrington (1978). "Hazardous solid waste from domestic wastewater treatment plants", *Environmental Health Perspectives*, 27, 231–237.

I. Haq, A.S. Kalamdhad (2021). "Phytotoxicity and cyto-genotoxicity evaluation of organic and inorganic pollutants containing petroleum refinery wastewater using plant bioassay", *Environmental Technology Innovation*, 23, 101651.

I. Haq, A.S. Kalmdhad (2023). "Enhanced biodegradation of toxic pollutants from paper industry wastewater using *Pseudomonas* sp. immobilized in composite biocarriers and its toxicity evaluation", *Bioresource Technology Reports*, 24, 101674.

I. Haq, A.S. Kalamdhad, A. Pandey (2022). "Genotoxicity evaluation of paper industry wastewater prior and post-treatment with laccase producing *Pseudomonas putida* MTCC 7525", *Journal of Cleaner Production*, 342, 130981.

I. Haq, S. Kumar, A. Raj, M. Lohani, G.N.V. Satyanarayana (2017). "Genotoxicity assessment of pulp and paper mill effluent before and after bacterial degradation using *Allium cepa* test", *Chemosphere*, 169, 642–650.

I. Haq, S. Kumar, V. Kumari, S.K. Singh, A. Raj (2016a). "Evaluation of bioremediation potentiality of ligninolytic *Serratia liquefaciens* for detoxification of pulp and paper mill effluent", *Journal of Hazard Materials*, 305, 190–199.

I. Haq, V. Kumari, S. Kumar, A. Raj, M. Lohani, R.N. Bhargava (2016b). "Evaluation of the phytotoxic and genotoxic potential of pulp and paper mill effluent Using *Vigna radiata* and *Allium cepa*", *Advanced Biology*, 2016, 1–10, Article ID 8065736.

I. Haq, A. Raj (2018). "Markandeya biodegradation of Azure-B dye by *Serratia liquefaciens* and its validation by phytotoxicity, genotoxicity and cytotoxicity studies", *Chemosphere*, 196, 58–68.

I. Ihsanullah, et al. (2020). "Bioremediation of dyes: Current status and prospects", *Journal of Water Process Engineering*, 38, 1–27.

I namuddin, M.I. Ahamed, et al. (2020). *Methods for Bioremediation of Water and Wastewater Pollution*, Environmental Chemistry for a sustainable world, vol. 51, 3–4. https://doi.org/10.1007/978-3-030-48985-4.

Y. Li, et al. (2009). "Water quality analysis of the Songhua river basin using multivariate techniques", *Journal of Water Resource. Prot.* 1, 110–121. https://doi.org/ 10.4236/jwarp.2009.12015.

D. R. Lovely, et al. (1997). "Bioremediation of contaminated metal", *Current Opinion in Biotechnology*, 8, 285–289.

S. Maryjoseph, et al. (2020). "Microalgae based wastewater treatment for the removal of emerging contaminants: A review of challenges and opportunities", *Case Studies in Chemical and Environmental Engineering*, 2, 1–10.

S. Muttamara (1996). "Wastewater characteristics", *Conservation and Recycling*, 16, 145–159.

F. A. Nasr, et al. (2004). "Chemical industry wastewater treatment", *Environmentalist*, 27, 275–286.

J. Nie, et al. (2018), "Bioremediation of water containing pesticides by microalgae: Mechanisms, methods, and prospects for future research", *Science of the Total Environment*, 707, 1–50.

N. Ojha, et al. (2021). "Bioremediation of industrial wastewater: A review", *Earth and Environmental Science*, 796, 1–30.

O. Paul, et al (2021). "In situ and ex situ bioremediation ff heavy metals: The present scenario", *Journal of Environmental Engineering and Landscape Management*, 29, 454–469.

C. V.T. Rigueto et al. (2020). "Alternative techniques for caffeine removal from wastewater: An overview of opportunities and challenges", *Journal of Water Process Engineering*, 35, 1–12.

P. Sharma, et al. (2021). "Critical review on microbial community during in-situ bioremediation of heavy metals from industrial wastewater", *Environmental Technology & Innovation*, 24, 1–16.

A. Sonune, et al. (2004). "Developments in wastewater treatment methods", *Desalination*, 167, 55–63.

D. L. Sutherland, et al. (2019). "Microalgal bioremediation of emerging contaminants - Opportunities and challenges", *Water Research*, 164, 1–13.

M. Tariq, et al. (2015). "Characteristics of industrial effluents and their possible impacts on quality of underground water", *Soil and Environment*, 25, 64–69.

S. M. Tripathi, et al. (2018). "Bioremediation of groundwater: An overview", *International Journal of Applied Engineering Research*, 13, 16825–16832.

A. F.M. Udaiyappan, et al. (2017). "A review of the potentials, challenges and current status of microalgae biomass applications in industrial wastewater treatment", *Journal of Water Process Engineering*, 20, 8–21.

M. Vidali (2001). "Bioremediation. An overview", *Pure and Applied Chemistry*, 73, 1163–1172.

G. S. Vishwakarma, et al. (2020). "Current status, challenges and future of bioremediation", *Bioremediation of Pollutants*, 20, 403–415.

Y. Wu, et al. (2014). "In situ bioremediation of surface waters by periphytons", *Bioresource Technology*, 151, 367–372.

W. Xiong, et al. (2019). "Biological consequences of environmental pollution in running water ecosystems: A case study in zooplankton". *Environmental Pollution*, 252, 1483–1490.

8 Microbial Community in a Wastewater System

Proteobacteria and Cyanobacteria

Shreya Anand and Padmini Padmanabhan

8.1 INTRODUCTION

Microalgae are a diverse group of prokaryotic and eukaryotic organisms that can be found in a variety of ecosystems with a variety of environments, including brackish, freshwater, marine, and freshwater environments (Renuka et al. 2018). The metabolic ability of microalgae to use inorganic and organic resources can be used to clean wastewater and produce biofuels (Abdo et al. 2016). Palmer divided the microalgae into 60 genera and 80 species based on their resistance to organic contaminants. More than 1000 taxa, including 725 species, 240 genera, and 125 variations, have been documented to have the potential to tolerate pollution. Eight green algae, six diatoms, six flagellates, and four blue-green algae are among the extremely tolerant microalgae genera (Palmer 1969).

Wastewater is a potentially useful, easily accessible, and affordable method for growing microalgae species (Ding et al. 2015). Wastewater contains microplastics, xenobiotics, and important components including nitrogen, carbon, and phosphorus that are crucial for the growth of microalgae. Using microalgae to grow in conjunction with wastewater treatment has various benefits, including the simultaneous intake of nutrients and the cost-effective generation of biomass and biofuels (Molazadeh et al. 2019). Scenedesmus, Chlamydomonas, and Chlorella cultivation in various wastewaters has demonstrated effective reduction of phosphorus and nitrogen (Gao et al. 2016). Furthermore, the skill to eliminate harmful molecules and metal may remove pollutants of subordinate pollution produced through different physico-chemical techniques (Matamoros et al. 2015). Growing microalgae in wastewater has the double benefit of removing components and producing biomass for the manufacture of biofuels. The combination of wastewater treatment and biofuel production has been the subject of a number of prior reports, but they all placed a strong emphasis on a single paradigm, biofuel production, genetic engineering or omics, and techno-economic analysis (El-sheekh et al. 2021).

A diverse collection of microorganisms known as cyanobacteria can be originated in a wide range of aquatic and terrestrial habitats (Svrcek & Smith 2004). They can undergo photosynthesis and create organic molecules because they are Gram-negative

DOI: 10.1201/9781003368120-8

eubacteria (Kirkwood et al. 2003), which sets cyanobacteria apart from other bacteria (Pitois et al. 2000). Most cyanobacteria are aerobic photoautotrophs since photosynthesis is the primary source of energy for them. Their metabolism depends on water, carbon dioxide, inorganic materials, light, and other elements. However, in their natural habitat, certain species can endure prolonged periods of darkness, and many cyanobacteria are capable of heterotrophic survival (Mur et al. 1999). Aquatic cyanobacteria thrive in certain environmental conditions, including high temperatures, nutrition availability, and light, which can result in cyanobacteria blooms.

Nutrient loadings from human activities are the primary source of eutrophication in all aquatic systems, with urban, industrial, and agricultural discharges playing a key role in the rise of phosphorus as well as nitrogen levels in getting streams (Bartram et al. 1999). When compared to the same levels seen in natural ecosystems, components in public and business discharges can be three levels higher extent (Kirkwood et al. 2003). Most waste waters should be treated to remove these nutrients and organic substances in order to mitigate these impacts (Haq and Kalamdhad 2021, 2023; Haq et al. 2016a, b, 2017, 2022; Haq and Raj 2018).

Wastewater treatment typically involves three stages: preliminary treatment, which primarily entails the settling and removal of solids; secondary treatment for the mineralization of the organic content; and tertiary treatment, which eliminates the levels of phosphorus and nitrogen as well as pathogenic bacteria (Crites et al. 2010). The majority of secondary treatment designs are open structures with microbial populations, primarily bacteria, which help to partially remove organic matter and nutrients before to discharge into receiving waters (Peavy et al. 1985). Accordingly, secondary treatment systems may provide the optimal environment for the growth of cyanobacteria, and numerous investigations have shown that these systems contain these organisms (Kirkwood et al. 2001; Vasconcelos & Pereira 2001).

On microalgal biofuel and its association in many domains, there are a few reports accessible. The evolutionary categorization of microalgae & cyanobacteria in relation to their potential for wastewater treatment is highlighted in this chapter. This chapter discusses about the diversity of the microbial communities w.r.t the microalgae and cyanobacteria; photosynthetic mechanism of microbial cells and the role of cyanobacteria and microalgae in the treatment of wastewater.

8.2 DIVERSITY OF MICROBIAL COMMUNITIES

The ability to consistently and accurately imitate full-scale effectiveness in reaction to reactor and process design, influent structure, and environmental conditions is necessary for the widespread and maintained implementation of microalgal and cyanobacterial treatment systems, even though technical and financial bottlenecks still exist. Lack of model accuracy and clarity regarding the model's structure and underlying research hinders this ability (Shoener et al. 2019).

It is possible for some microalgae and cyanobacteria to use mixotrophic, phototrophic, and heterotrophic metabolisms (Adesanya et al. 2014; Zhang et al. 1998; Lopo et al. 2012). The metabolism in use is influenced by external factors including substrate accessibility and illumination. Additionally, carbon absorption and partitioning can be impacted by the presence or lack of nutrients. These intricate processes are usually represented by models that either I have more variables than most versions of heterotrophic bacteria do, or (ii) make false simplifications that reduce the model's accuracy. There are hundreds of simulations for algae as a result of these divergent methods, which shows that this research lacks a clear direction.

The goal of early modeling attempts was to comprehend the behavior of phytoplankton in natural ecosystems, however applying empirically obtained models from nature to engineered systems necessitates verification and may even require modification (Jørgensen 1976; Steele 1962). The development of generalizable model structures and precisely defined parameters pertinent to WRRFs has also been hampered by divergent approaches to modeling cyanobacterial and algal processes, highly variable experimental conditions, and a disregard for already existing chemotrophic model structures.

The amount of nutrients that must be removed by water resource recovery facilities (WRRFs) is getting close to the capacity of available technology. The development of latest technologies capable of reliably extracting all types of nutrients, which include dissolved organic nitrogen and phosphorus, is necessary to remove nitrogen & phosphorus beyond the current technological limit as untreated sewage requirements are becoming more stringent (Bott and Parker 2011). By simultaneously reaching effluent levels of nitrogen and phosphorus below the existing technological limit and enabling nutrient reuse, microalgal resource recovery devices could greatly enhance nutrient use efficiency of wastewaters (Leow et al. 2015; Metting 1996).

The variety of models available to simulate algal development are discussed, but there is no clear direction as to when subcomponents should be taken into account or left out, or as to why a certain equation should be used to mimic each subcomponent (Darvehei et al. 2018; Lee et al. 2015). Recent process models have made an effort to balance these disagreements, but there is still a need for an industry-wide, harmonized consensus (Baroukh et al. 2014; Guest et al. 2013; Wagner et al. 2016). Establishing a consistent modeling framework that can take into consideration pertinent process and environmental parameters while avoiding superfluous complexity is essential to advancing the wider deployment of algae and cyanobacteria process models by academics and practitioners.

Different evolutionary lineages are used to categorize microalgae. Photoautotrophic microalgae perform photosynthesis in chloroplast (Salama et al. 2013). The most important traits are the extreme diversity in morphology, physiology, and gene content. The Chromalveolata, Archaeplastida, Excavata, and Rhizaria subgroups make up four of the six major eukaryotic groups in which microalgae are found. When compared to Archaeplastida, the majority of these subgroups' microalgae have a strong relationship with Chromalveolata (Salama et al. 2019a). Based on molecular clocks, a polyphyletic group of >5,000 species of red algae, <16,000 species of green algae, <3,000 species of dinophyceans, >15,000 species of chromophytes, around 900 species of euglenophytes, 200 species of cryptophytes, and 13 species of glaucophytes has evolved independently over geological time at varying rates (Andersen 1992).

Cyanobacteria range from filamentous to colony bacteria and can divide through binary or multiple fission. Taxonomically, binary fission and multiple fission are used to categorize unicellular organisms. The non-heterocystous filamentous forms distinguish between branching filaments that have undergone numerous fissions and heterocysts that do not. Gloeobacterales genus Gloeobacter, which was formerly classed under the Chroococcales, was proposed as the sixth cyanobacterial order (Cavalier-Smith 2002; Demoulin et al. 2019).

8.3 MECHANISM OF PROTEOBACTERIAL AND CYANOBACTERIAL CELLS

Microalga and cyanobacteria both engage in photosynthesis, making them the primary organisms of interest for the production of biofuels (Abomohra et al. 2019). Together have rapid rates of growth without needing a lot of arable land. Growing under submerged water enables higher carbon dioxide use in wastewater than in ambient air. Light and carbon dioxide from the atmosphere are taken in by microalgae and later converted into chemical energy. Carbohydrates, lipids, and proteins in the cell serve as storage for this chemical energy. Microalgae grow faster and have a higher photosynthetic rate, giving them various advantages over terrestrial plants (Sharma et al. 2018).

Understanding the electron transport mechanism connected to the source of organic and the conversion of light energy into chemical energy is essential in the context of biorefineries. Membrane proteins and photosystem I and II (PS I and II) make up the linear route for electron transport. In this system of linear electron flow, the splitting of the water molecule into protons, oxygen, and electrons occurs initially through the excitation of PS II by light energy. ATP is produced concurrently with the simultaneous delivery of protons to the thylakoid lumen during this electron transfer process (Engin et al. 2018). Additionally, a process known as mixotrophic metabolism allows some microalgae species to efficiently use both organic and inorganic carbon at the same time when exposed to light (Khan et al. 2018).

In phototrophic cultivation, open ponds and closed photobioreactors are typically used as two different systems for the growth of microalgae. When it comes to producing microalgal species that can survive in harsh settings like high salinity and high pH, closed PBR is preferable to open PBR systems, especially in the case of microalgal species, in order to prevent contamination by other bacteria (Nagappan et al. 2019; Khan et al. 2017). Both cyanobacteria and microalgae use a similar photosynthesis mechanism. Cyanobacteria have bilin photoreceptors, which allow them to detect UV and visible light. These photoreceptors' main duties include controlling salt acclimatization, state transition, and both positive and negative phototaxis. They are critical for carotenoid accumulation and filament stacking, respectively. In order to discriminate between different domain architectures and cysteine amino acid residues, which is necessary for the attachment of tetrapyrrole chromophores (Wiltbank and Kehoe 2016).

8.4 ROLE OF PROTEOBACTERIA AND CYANOBACTERIA IN THE REMOVAL OF RECALCITRANT POLLUTANTS FROM WASTEWATER

Due to their high application costs, conventional techniques for removing pollutants from wastewater streams fall short of meeting the existing regulatory discharge limitations (Salama et al. 2019b). The desire for affordable alternative technologies has increased as a result of these limitations. Coagulation-flocculation, chemical precipitation, membrane filtration, Ion exchange, oxidation with hydrogen peroxide, adsorption, electrochemical techniques, etc. are a few examples of the physico-chemical approaches that have been extensively studied for wastewater treatment (Zouboulis et al. 2015).

The screening process for treatment approaches are mostly based on a variety of variables. These methods' main limitations are their high energy consumption, instability of the processes, and length of time, as well as high carbon emission, excessive sludge discharge, and resource waste. Additionally, the idea of ecological development in wastewater treatment with minimal carbon emissions, low energy usage, & resource recycling is clearly hampered by these approaches (Sun et al. 2016). Because wastewater contains complex components, these processes alone are unable to produce water that meets the recommended standards. Many different approaches have been combined to get beyond the financial obstacle (Olguin and Sanchez-Galvan 2012).

By metabolizing and eliminating various pollutants, the use of microalgae in wastewater treatment is an efficient method for water bioremediation (Salama et al. 2017). Growing microalgae efficiently traps and breaks down resistant organic molecules to increase biomass production. Following harvest, the generated biomass can be successfully converted into a variety of useful bioproducts that have been found to excel in terms of life cycle assessment (LCA) impact (Woertz et al. 2014).

The ability to produce microalgae while simultaneously treating wastewater depends on the ability to use nutrients and organic matter found in wastewater to fuel either heterotrophic or mixotrophic growth (Salama et al. 2017). Theoretically, the process is explained as the result of microalgae turning wastewater's dissolved organic carbon into carbon dioxide, which they then recoup through photosynthesis (Gupta and Pawar 2018). About 75 years ago, the first commercial strains of microalgae and cyanobacteria were produced, and they were used to treat wastewater.

Applications of phytoremediation include: (i) removing nutrients from municipal wastewater and organically significant effluents; (ii) using algal biosorbents to remove xenobiotic compounds and nutrients; (iii) treating acidic and metal-containing wastewater; (iv) sequestering carbon dioxide; and (v) using microalgae-based biosensors to detect toxic compounds. By reacting to different poisons and metal ions, microalgae proliferation indicates water contamination. Blue-green algae are unquestionably ideal for the twin function of treating wastewater while effectively utilizing the numerous chemicals necessary for development and increasing biomass production (Shahid et al. 2020).

It was suggested that cyanobacteria can be used to remediate wastewater, and pilot tests were carried out (Caldwell 1946; Oswald and Gotaas 1957). Due to the clear potential to adopt inorganics, the motivation for using cyanobacteria in wastewater treatment has developed since the beginning of 1980. Because of the tremendous flexibility of its metabolic machinery, cyanobacteria offer an intriguing alternative that can be applied to bioremediation in this situation. When compared to fungus and bacteria, mass cyanobacterial cultivation is more efficient and less expensive. A current requirement is the manipulation of cyanobacteria in wastewater treatment.

As a promising alternative to traditional biological processes, wastewater treatment systems utilizing photosynthetic bacteria have recently emerged (Gonçalves et al. 2017). However, over the past ten years, there has been a significant increase in research interest in the use of cyanobacteria to treat wastewater discharged from various locations, and this work has produced encouraging results for both organic and inorganic discharge (Subramaniyam et al. 2016; Ferreira et al. 2017; Papadopoulos et al. 2020a, b). Reduced operating costs due to photosynthesis replacing mechanical aeration requirements, reduced carbon dioxide emissions, and a reduction in the high expense of employing chemical fertilizers for microalgal development are just a few of the advantages of this technique (Ghimire et al. 2017). Additionally, the accumulating biomass can be used as a resource to create biofuels (Saratale et al. 2018; Koutra et al. 2018). Additionally, cyanobacterial biofilms show significant promise for use in the treatment of wastewater. The biofilm matrix protects microbial residents from external environmental stressors (Liu et al. 2017).

Cyanobacteria can be producers instead of consumers thanks to their photoautotrophic nature and some species' ability to fix atmospheric nitrogen, which lowers the cost of their production. Additionally, cyanobacteria are known to proliferate in great numbers in contaminated environments, where they are extensively dispersed and predominate the microfloral communities (Whitton 1992). Because of their presence in the polluted systems, they develop innate resistance and selectivity against environmental toxins. However, lowering levels of biodegradable contaminants does not enhance their sustainability or metabolic activities. They have been successfully employed as a low-cost remediation technique for sewage effluents, dissolved inorganic nutrients from fish farms, and dairy wastewater (Lincoln et al. 1996; Hu et al. 2000; Phang et al. 2000; de-Bashan et al. 2004). With the goal of standardizing their beneficial use for wastewater reclamation, many cyanobacterial strains have been thoroughly sequenced and investigated.

Aquatic cyanobacteria thrive in certain environmental conditions, including high temperatures, nutrition availability, and light, which can result in cyanobacteria blooms. Nutrient loading from human activities is the primary source of eutrophication in all aquatic systems, with urban, industrial, and agricultural discharges playing a key role in the rise of nitrogen and phosphorus levels in receiving streams (Bartram et al. 1999). When compared to the same levels seen in natural ecosystems, nutrients in municipal and industrial discharges can be three orders of magnitude higher (Kirkwood et al. 2003). Most waste waters should be treated to remove these nutrients and organic substances in order to mitigate these impacts.

Eutrophication is caused by the occurrence of nutritional components, in the form of nitrate, nitrite, ammonia, in wastewater (Liu et al. 2010). With a wide range of variation in their organization and habitat, cyanobacteria produce a diverse set of organisms, some of which have the characteristic of photoautotrophic metabolism and are found in both freshwater and marine environments (Lee 2008). A major component of the microbial variety in wastewaters is cyanobacteria, which can also contribute to the wastewaters' ability to purify themselves (Sen et al. 2013). Its capacity for adaptation to severe and challenging settings has enabled the scientific community to screen, identify suitable strains, and create effective wastewater treatment methods (Fouilland 2012).

To regulate the techno-economic assessment of microalgal productions to expose its economic, social, energy, and environmental implications as well as associated issues, several research groups have made an effort (Yadav et al. 2021). The presence of turbidity and total solids in wastewater is one of the restrictions on the growth of microalgae and cyanobacteria in wastewater. If untreated wastewater is used, high total solids (TS) and turbidity prevent the light from passing through during microalgal growth. High organic carbon concentration in wastewater, which promotes the growth of rival species and consequently impacts the microalgal growth rate, is another factor that limits the growth of microalgae. Although there is no defined modeling framework, cyanobacterial technologies have the potential to produce effluent nutrient concentrations below those of current biological nutrient removal systems.

8.5 SUMMARY AND CONCLUSION

A variety of microalgae and cyanobacteria are excellent candidates for biomass generation in conjunction with wastewater treatment due to their high photosynthetic efficiency, rapid development, and resistance to wastewater toxicity. Growing microalgae in a variety of wastewater can remove metal, phosphate, and nitrogen efficiently. The main removal processes used by the microalgae are bioaccumulation and bioadsorption, which, in comparison to conventional techniques, are economical and environmentally friendly. To boost microalgae cultivation and reduce growth-inhibiting wastewater characteristics, several operational tactics and multidisciplinary integration have been implemented. Omics and metabolic engineering could be used to improve biotechnological applications even more.

Numerous research show that cyanobacteria have powerful uses in the treatment of wastewater, and their importance has long been recognized. Cyanobacteria are suitable candidates for cleaning wastewater because of their photosynthetic and autotrophic modes of nutrition. This gives them an added advantage since, in addition to treating the wastewater, they may also replenish the waterways. Before they may be fully utilized in water treatment facilities, more thorough and extensive field trials are necessary. Another issue that must be addressed to prevent eutrophication is the separation of cyanobacterial biomass from treated water flow. A better business outlook may result from cyanobacterial biomass production combined with biological wastewater treatment.

In several investigations, mixed-culture cyanobacterial consortiums have produced superior outcomes than single-culture methods, it has been observed. However, it was sometimes seen that the axenic culture performed better than the mixed cultures. This may be because the mixed cultures compete with one another for nutrients. In order to prepare reusable immobilized particles containing cyanobacteria for creating a system that is cost-effective, more study is needed to address these problems. In order to do this, it is necessary to investigate suitable immobilization techniques, such as co-immobilization of different species, which will enable them to engage symbiotically and boost removal capacities.

REFERENCES

Abdo, S. M., El-Enin, S. A., El-Khatib, K. M., El-Galad, M. I., Wahba, S. Z., El Diwani, G., & Ali, G. H. (2016). Preliminary economic assessment of biofuel production from microalgae. *Renewable and Sustainable Energy Reviews, 55*, 1147–1153.

Abomohra, A. E. F., Shang, H., El-Sheekh, M., Eladel, H., Ebaid, R., Wang, S., & Wang, Q. (2019). Night illumination using monochromatic light-emitting diodes for enhanced microalgal growth and biodiesel production. *Bioresource Technology, 288*, 121514.

Adesanya, V. O., Davey, M. P., Scott, S. A., & Smith, A. G. (2014). Kinetic modelling of growth and storage molecule production in microalgae under mixotrophic and autotrophic conditions. *Bioresource Technology, 157*, 293–304.

Andersen, R. A. (1992). Diversity of eukaryotic algae. *Biodiversity & Conservation, 1*, 267–292.

Baroukh, C., Muñoz-Tamayo, R., Steyer, J. P., & Bernard, O. (2014). DRUM: A new framework for metabolic modeling under non-balanced growth. Application to the carbon metabolism of unicellular microalgae. *PLoS One, 9*(8), e104499.

Bartram, J., Carmichael, W. W., Chorus, I., Jones, G., & Skulberg, O. M. (1999). Chapter 1. Introduction. In *Toxic Cyanobacteria in Water: A Guide to Their Public Health Consequences, Monitoring and Management*, pp. 352–358. https://doi.org/10.1201/9781003081449-1.

Bott, C. B., & Parker, D. S. (2011). *WEF/WERF Study Quantifying Nutrient Removal Technology Performance* (Vol. 10). Alexandria, VA: Water Environment Research Foundation.

Caldwell, D. H. (1946). Sewage oxidation ponds: Performance, operation and design. *Sewage Works Journal, 18*(3), 433–458.

Crites, R. W., Middlebrooks, E. J., & Reed, S. C. (2010). *Natural Wastewater Treatment Systems*. Boca Raton, FL: CRC Press.

Cavalier-Smith, T. (2002). The phagotrophic origin of eukaryotes and phylogenetic classification of Protozoa. *International Journal of Systematic and Evolutionary Microbiology, 52*(2), 297–354.

De-Bashan, L. E., Hernandez, J. P., Morey, T., & Bashan, Y. (2004). Microalgae growth-promoting bacteria as "helpers" for microalgae: A novel approach for removing ammonium and phosphorus from municipal wastewater. *Water Research, 38*(2), 466–474.

Darvehei, P., Bahri, P. A., &Moheimani, N. R. (2018). Model development for the growth of microalgae: A review. *Renewable and Sustainable Energy Reviews, 97*, 233–258.

Demoulin, C. F., Lara, Y. J., Cornet, L., François, C., Baurain, D., Wilmotte, A., & Javaux, E. J. (2019). Cyanobacteria evolution: Insight from the fossil record. *Free Radical Biology and Medicine, 140*, 206–223.

Ding, J., Zhao, F., Cao, Y., Xing, L., Liu, W., Mei, S., & Li, S. (2015). Cultivation of microalgae in dairy farm wastewater without sterilization. *International Journal of Phytoremediation, 17*(3), 222–227.

El-Sheekh, M., El-Dalatony, M. M., Thakur, N., Zheng, Y., &Salama, E. S. (2021). Role of microalgae and cyanobacteria in wastewater treatment: Genetic engineering and omics approaches. *International Journal of Environmental Science and Technology, 19*, 1–22.

Engin, I. K., Cekmecelioglu, D., Yücel, A. M., & Oktem, H. A. (2018). Evaluation of heterotrophic and mixotrophic cultivation of novel *Micractinium* sp. ME05 on vinasse and its scale up for biodiesel production. *Bioresource Technology, 251*, 128–134.

Ferreira, A., Ribeiro, B., Marques, P. A., Ferreira, A. F., Dias, A. P., Pinheiro, H. M., … & Gouveia, L. (2017). *Scenedesmus obliquus* mediated brewery wastewater remediation and CO$_2$ biofixation for green energy purposes. *Journal of Cleaner Production, 165*, 1316–1327.

Fouilland, E. (2012). Biodiversity as a tool for waste phycoremediation and biomass production. *Reviews in Environmental Science and Bio/Technology, 11*(1), 1–4.

Gao, F., Li, C., Yang, Z. H., Zeng, G. M., Mu, J., Liu, M., & Cui, W. (2016). Removal of nutrients, organic matter, and metal from domestic secondary effluent through microalgae cultivation in a membrane photobioreactor. *Journal of Chemical Technology & Biotechnology*, *91*(10), 2713–2719.

Ghimire, A., Kumar, G., Sivagurunathan, P., Shobana, S., Saratale, G. D., Kim, H. W., ... & Munoz, R. (2017). Bio-hythane production from microalgae biomass: Key challenges and potential opportunities for algal bio-refineries. *Bioresource Technology*, *241*, 525–536.

Gonçalves, A. L., Pires, J. C., & Simões, M. (2017). A review on the use of microalgal consortia for wastewater treatment. *Algal Research*, *24*, 403–415.

Guest, J. S., Van Loosdrecht, M. C., Skerlos, S. J., & Love, N. G. (2013). Lumped pathway metabolic model of organic carbon accumulation and mobilization by the alga *Chlamydomonas reinhardtii*. *Environmental Science & Technology*, *47*(7), 3258–3267.

Gupta, S., & Pawar, S. B. (2018). An integrated approach for microalgae cultivation using raw and anaerobic digested wastewaters from food processing industry. *Bioresource Technology*, *269*, 571–576.

Haq, I., & Kalmdhad, A.S. (2023). Enhanced biodegradation of toxic pollutants from paper industry wastewater using Pseudomonas sp. immobilized in composite biocarriers and its toxicity evaluation. *Bioresource Technology Report*, 24, 101674.

Haq, I., Kalamdhad, A.S. (2021). Phytotoxicity and cyto-genotoxicity evaluation of organic and inorganic pollutants containing petroleum refinery wastewater using plant bioassay. *Environmental Technology Innovation*, 23, Article 101651.

Haq, I., Kalamdhad, A.S., Pandey, A. (2022). Genotoxicity evaluation of paper industry wastewater prior and post-treatment with laccase producing Pseudomonas putida MTCC 7525. *Journal of Cleaner Production*, 342, Article 130981.

Haq, I., Kumar, S., Kumari, V., Singh, S.K., Raj, A. (2016a). Evaluation of bioremediation potentiality of ligninolytic *Serratia liquefaciens* for detoxification of pulp and paper mill effluent. *Journal of Hazardous Materials*, 305, 190–199.

Haq, I., Kumar, S., Raj, A., Lohani, M., Satyanarayana, G.N.V. (2017). Genotoxicity assessment of pulp and paper mill effluent before and after bacterial degradation using Allium cepa test. *Chemosphere*, 169, 642–650.

Haq, I., Kumari, V., Kumar, S., Raj, A., Lohani, M., Bhargava, R.N. (2016b). Evaluation of the phytotoxic and genotoxic potential of pulp and paper mill effluent Using Vigna radiata and Allium cepa. *Advanced Biology*, 2016, 1–10, Article ID 8065736.

Haq, I., & Raj, A. (2018). Markandeya Biodegradation of Azure-B dye by Serratia liquefaciens and its validation by phytotoxicity, genotoxicity and cytotoxicity studies. *Chemosphere*, 196, 58–68.

Hu, Q., Westerhoff, P., & Vermaas, W. (2000). Removal of nitrate from groundwater by cyanobacteria: Quantitative assessment of factors influencing nitrate uptake. *Applied and Environmental Microbiology*, *66*(1), 133–139.

Jørgensen, S. E. (1976). A eutrophication model for a lake. *Ecological Modelling*, *2*(2), 147–165.

Khan, M. I., Shin, J. H., & Kim, J. D. (2018). The promising future of microalgae: current status, challenges, and optimization of a sustainable and renewable industry for biofuels, feed, and other products. *Microbial Cell Factories*, *17*(1), 1–21.

Khan, S., Siddique, R., Sajjad, W., Nabi, G., Hayat, K. M., Duan, P., & Yao, L. (2017). Biodiesel production from algae to overcome the energy crisis. *HAYATI Journal of Biosciences*, *24*(4), 163–167.

Kirkwood, A. E., Nalewajko, C., & Fulthorpe, R. R. (2003). Physiological characteristics of cyanobacteria in pulp and paper waste-treatment systems. *Journal of Applied Phycology*, *15*, 325–335.

Koutra, E., Economou, C. N., Tsafrakidou, P., & Kornaros, M. (2018). Bio-based products from microalgae cultivated in digestates. *Trends in Biotechnology*, *36*(8), 819–833.

Lee, E., Jalalizadeh, M., & Zhang, Q. (2015). Growth kinetic models for microalgae cultivation: A review. *Algal Research*, *12*, 497–512.

Lee, R. E. (2018). *Phycology*. Cambridge: Cambridge University Press.

Leow, S., Witter, J. R., Vardon, D. R., Sharma, B. K., Guest, J. S., & Strathmann, T. J. (2015). Prediction of microalgae hydrothermal liquefaction products from feedstock biochemical composition. *Green Chemistry*, *17*(6), 3584–3599.

Liu, J., Wu, Y., Wu, C., Muylaert, K., Vyverman, W., Yu, H. Q., … & Rittmann, B. (2017). Advanced nutrient removal from surface water by a consortium of attached microalgae and bacteria: A review. *Bioresource Technology*, *241*, 1127–1137.

Liu, W., Zhang, Q., & Liu, G. (2010). Lake eutrophication associated with geographic location, lake morphology and climate in China. *Hydrobiologia*, *644*, 289–299.

Lincoln, E. P., Wilkie, A. C., & French, B. T. (1996). Cyanobacterial process for renovating dairy wastewater. *Biomass and Bioenergy*, *10*(1), 63–68.

Lopo, M., Montagud, A., Navarro, E., Cunha, I., Zille, A., De Córdoba, P. F., … & Urchueguia, J. F. (2012). Experimental and modeling analysis of Synechocystis sp. PCC 6803 growth. *Microbial Physiology*, *22*(2), 71–82.

Matamoros, V., Gutiérrez, R., Ferrer, I., García, J., & Bayona, J. M. (2015). Capability of microalgae-based wastewater treatment systems to remove emerging organic contaminants: a pilot-scale study. *Journal of Hazardous Materials*, *288*, 34–42.

Molazadeh, M., Ahmadzadeh, H., Pourianfar, H. R., Lyon, S., & Rampelotto, P. H. (2019). The use of microalgae for coupling wastewater treatment with CO_2 biofixation. *Frontiers in Bioengineering and Biotechnology*, *7*, 42.

Metting, F. B. (1996). Biodiversity and application of microalgae. *Journal of Industrial Microbiology*, *17*, 477–489.

Nagappan, S., Devendran, S., Tsai, P. C., Dahms, H. U., & Ponnusamy, V. K. (2019). Potential of two-stage cultivation in microalgae biofuel production. *Fuel*, *252*, 339–349.

Olguín, E. J., & Sánchez-Galván, G. (2012). Heavy metal removal in phytofiltration and phycoremediation: The need to differentiate between bioadsorption and bioaccumulation. *New Biotechnology*, *30*(1), 3–8.

Oswald, W. J., & Gotaas, H. B. (1957). Photosynthesis in sewage treatment. *Transactions of the American Society of Civil Engineers*, *122*(1), 73–97.

Palmer, C. M. (1969). A composite rating of algae tolerating organic pollution 2. *Journal of Phycology*, *5*(1), 78–82.

Papadopoulos, K. P., Economou, C. N., Dailianis, S., Charalampous, N., Stefanidou, N., Moustaka-Gouni, M., … & Vayenas, D. V. (2020a). Brewery wastewater treatment using cyanobacterial-bacterial settleable aggregates. *Algal Research*, *49*, 101957.

Papadopoulos, K. P., Economou, C. N., Tekerlekopoulou, A. G., & Vayenas, D. V. (2020b). Two-step treatment of brewery wastewater using electrocoagulation and cyanobacteria-based cultivation. *Journal of Environmental Management*, *265*, 110543.

Peavy, H. S., Rowe, D. R., & Tchobanoglous, G. (1985). *Environmental Engineering* (Vol. 2985). New York: McGraw-Hill.

Phang, S. M., Miah, M. S., Yeoh, B. G., & Hashim, M. A. (2000). Spirulina cultivation in digested sago starch factory wastewater. *Journal of Applied Phycology*, *12*, 395–400.

Pitois, S., Jackson, M. H., & Wood, B. J. B. (2000). Problems associated with the presence of cyanobacteria in recreational and drinking waters. *International Journal of Environmental Health Research*, *10*(3), 203–218.

Renuka, N., Guldhe, A., Prasanna, R., Singh, P., & Bux, F. (2018). Microalgae as multifunctional options in modern agriculture: Current trends, prospects and challenges. *Biotechnology Advances*, *36*(4), 1255–1273.

Salama, E. S., Govindwar, S. P., Khandare, R. V., Roh, H. S., Jeon, B. H., & Li, X. (2019a). Can omics approaches improve microalgal biofuels under abiotic stress? *Trends in Plant Science*, *24*(7), 611–624.

Salama, E. S., Kim, H. C., Abou-Shanab, R. A., Ji, M. K., Oh, Y. K., Kim, S. H., & Jeon, B. H. (2013). Biomass, lipid content, and fatty acid composition of freshwater Chlamydomonasmexicana and Scenedesmusobliquus grown under salt stress. *Bioprocess and Biosystems Engineering*, *36*, 827–833.

Salama, E. S., Kurade, M. B., Abou-Shanab, R. A., El-Dalatony, M. M., Yang, I. S., Min, B., & Jeon, B. H. (2017). Recent progress in microalgal biomass production coupled with wastewater treatment for biofuel generation. *Renewable and Sustainable Energy Reviews*, *79*, 1189–1211.

Salama, E. S., Roh, H. S., Dev, S., Khan, M. A., Abou-Shanab, R. A., Chang, S. W., & Jeon, B. H. (2019b). Algae as a green technology for heavy metals removal from various wastewater. *World Journal of Microbiology and Biotechnology*, *35*, 1–19.

Saratale, R. G., Kumar, G., Banu, R., Xia, A., Periyasamy, S., & Saratale, G. D. (2018). A critical review on anaerobic digestion of microalgae and macroalgae and co-digestion of biomass for enhanced methane generation. *Bioresource Technology*, *262*, 319–332. https://doi.org/10.1016/j.biortech.2018.03.030

Sen, B., Alp, M. T., Sonmez, F., Kocer, M. A. T., & Canpolat, O. (2013). Relationship of algae to water pollution and waste water treatment. *Water Treatment*, *14*, 335–354.

Shahid, A., Malik, S., Zhu, H., Xu, J., Nawaz, M. Z., Nawaz, S., …& Mehmood, M. A. (2020). Cultivating microalgae in wastewater for biomass production, pollutant removal, and atmospheric carbon mitigation: A review. *Science of the Total Environment*, *704*, 135303.

Sharma, P. K., Saharia, M., Srivstava, R., Kumar, S., & Sahoo, L. (2018). Tailoring microalgae for efficient biofuel production. *Frontiers in Marine Science*, *5*, 382.

Sun, Y., Chen, Z., Wu, G., Wu, Q., Zhang, F., Niu, Z., & Hu, H. Y. (2016). Characteristics of water quality of municipal wastewater treatment plants in China: implications for resources utilization and management. *Journal of Cleaner Production*, *131*, 1–9.

Shoener, B. D., Schramm, S. M., Béline, F., Bernard, O., Martínez, C., Plósz, B. G., …& Guest, J. S. (2019). Microalgae and cyanobacteria modeling in water resource recovery facilities: A critical review. *Water Research X*, *2*, 100024.

Steele, J. H. (1962). Environmental control of photosynthesis in the sea. *Limnology and Oceanography*, *7*(2), 137–150.

Subramaniyam, V., Subashchandrabose, S. R., Ganeshkumar, V., Thavamani, P., Chen, Z., Naidu, R., & Megharaj, M. (2016). Cultivation of Chlorella on brewery wastewater and nano-particle biosynthesis by its biomass. *Bioresource Technology*, *211*, 698–703.

Svrcek, C., & Smith, D. W. (2004). Cyanobacteria toxins and the current state of knowledge on water treatment options: A review. *Journal of Environmental Engineering and Science*, *3*(3), 155–185.

Vasconcelos, V. M., & Pereira, E. (2001). Cyanobacteria diversity and toxicity in a wastewater treatment plant (Portugal). *Water Research*, *35*(5), 1354–1357.

Wágner, D. S., Valverde-Pérez, B., Sæbø, M. Bregua de la Sotilla, M., Van Wagenen, J., Smets, BF, & Plósz, BG, 2016. Towards a consensus-based biokinetic model for green microalgae-The ASM-A. *Water Res*, *103*, 485–499.

Wiltbank, L. B., & Kehoe, D. M. (2016). Two cyanobacterial photoreceptors regulate photosynthetic light harvesting by sensing teal, green, yellow, and red light. *MBio*, *7*(1), e02130–15.

Whitton, B. A. (1992). Diversity, ecology, and taxonomy of the cyanobacteria. *Photosynthetic Prokaryotes*, 1–51. https://doi.org/10.1007/978-1-4757-1332-9_1.

Woertz, I. C., Benemann, J. R., Du, N., Unnasch, S., Mendola, D., Mitchell, B. G., & Lundquist, T. J. (2014). Life cycle GHG emissions from microalgal biodiesel – A CA-GREET model. *Environmental Science & Technology*, *48*(11), 6060–6068. https://doi.org/10.1021/es403768q

Yadav, G., Shanmugam, S., Sivaramakrishnan, R., Kumar, D., Mathimani, T., Brindhadevi, K., … & Rajendran, K. (2021). Mechanism and challenges behind algae as a wastewater treatment choice for bioenergy production and beyond. *Fuel*, *285*, 119093.

Zhang, X. W., Zhang, Y. M., & Chen, F. (1998). Kinetic models for phycocyanin production by high cell density mixotrophic culture of the microalga *Spirulina platensis*. *Journal of Industrial Microbiology and Biotechnology*, *21*, 283–288.

Zouboulis, A. I., Peleka, E. N., & Samaras, P. (2015). Removal of toxic materials from aqueous streams. In *Mineral Scales and Deposits*, pp. 443–473. Elsevier. https://doi.org/10.1016/B978-0-444-63228-9.00017-6.

9 Potentials of Microbes in Wastewater Treatment and Management

Maghimaa Mathanmohun, Boojhana Elango, Gunadhor Singh Okram, and Maulin P. Shah

9.1 INTRODUCTION

The availability of clean water has a profound impact on human development, as it is considered a basic human right that remains inaccessible to a significant portion of the global population. By 2030, the urban population is projected to double, surpassing 5 billion, and if left unaddressed, water pollution issues will escalate, adversely affecting public health. Therefore, it is crucial that as a country's economy grows, a portion of their wealth should be dedicated to improving sanitation infrastructure. One prominent concern dominating international media headlines is the quantity and quality of waste being discharged into natural water bodies, prompting wealthy nations to allocate substantial resources to mitigate pollution among large populations, as observed by Oh et al. (2010). Additionally, reports of global contamination of underground water sources with heavy metals have emerged as a significant public health concern. To address these issues effectively, a comprehensive understanding of wastewater sources, encompassing biological, chemical, and physical aspects, as highlighted by Ferrera and Sanchez (2016), becomes imperative.

Therefore, throughout the world with the betterment of research facilities have led to the development and continual enhancement of diverse water treatment methods. These methods are categorized on the nature of wastewater source, resulting in the classification of water treatment technologies into physical, chemical, and biological treatment techniques, with further distinctions between ex situ and in situ technologies. To delve deeper into the applications of ex situ and in situ technologies, the in situ remediation actions are characterized by onsite treatment, while the elimination of pollutants at a distant location typifies ex situ technology, as explained by Ng et al. (2016). An illustrative instance of chemical water treatment is Flocculation which is used for in situ treatment of surface water plus groundwater. Water diversion is one of the examples of physical water treatments. However, due to their negative side effects, these two modes of water treatment could prove to be less popular. Flocculation, for instance, requires careful handling of chemicals in large quantities, posing potential hazards to users, while water diversion can incur substantial costs if implemented on a large scale, as cautioned by Della Rocca et al. (2007).

DOI: 10.1201/9781003368120-9

153

Last but not least perhaps, biological wastewater treatment technology would be the best treatment technique. 'Biodegradation' is the term that might be a synonym for biological wastewater treatment technology. In the effluent treatment method, many types of microorganisms are present. Conversely, the best-suited microorganism to the "environment" or conditions in the system will be the ones that will dominate under the situation. A certain type of microorganism is suited for wastewater treatment systems and is designed to promote an "environment". These microbes filter organic wastes from water and also for easy removal they "settle out" as solid material (Shaikh et al. 2013). For the precise type of microorganism wastewater treatment, operators are necessary to retain accurate conditions in the treatment system. Through the combined activity of microorganisms, including fungi, bacteria, algae, rotifers and protozoa, biological degradation of organic wastes are accomplished (Haq and Kalamdhad, 2021, 2023; Haq et al., 2016a, b, 2017, 2022; Haq and Raj, 2018). Biodegradation is the process by which organic molecules are broken down by microorganisms giving rise to the formation of carbon dioxide and water/methane (Nasr et al. 2014). For degrading pollutants and maintaining the stabilization of biological systems, microorganisms are the possible key players. Microorganism plays a pivotal part in wastewater treatment, contributing significantly to various biological treatment processes. Their functions encompass the putrefaction of organic matter, the coagulation of non-settleable colloidal solids, the removal of carbonaceous biochemical oxygen demand (BOD), and the stabilization of organic matter. Moreover, microorganisms are instrumental in converting colloids and dissolved carbonaceous organic matter into gaseous byproducts. Additionally, they make use of organic pollutants as sources of nitrogen, energy and carbon for their growth. Biological wastewater treatment technologies offer several advantages over alternative methods. They are comparatively cost-effective, typically result in minimal or no secondary release of pollutants, and notably exhibit lower detrimental impacts on the environment, as highlighted by Mingjun et al. (2009). Furthermore, when considering maintenance costs and capital investments, biological wastewater treatment technology holds an economic edge over both chemical and physical treatment approaches, as elucidated by Mittal (2011). This paper therefore attempts to give a crystal-clear explanation on the potentiality of microbes in wastewater treatment and their applications and advantages to mankind.

9.2   SOURCES OF WASTEWATER

Wastewater can be broadly categorized into four main types: industrial, domestic, urban and agricultural. In urban areas, wastewater consists of domestic sewage, industrial effluent, rainwater, and infiltration from surrounding sewage systems. In rural areas, agricultural wastewater primarily originates from agricultural activities, farms and occasionally contaminated groundwater, as noted by Hamdy et al. (2005). It's worth noting that domestic and industrial sewage are also contributors to contamination. Agricultural runoff is rich in inorganic nutrients like nitrogen (N), phosphorus (P), and toxic chemicals, which can lead to eutrophication of surface water. Domestic wastewater, primarily composed of animal and human waste, includes contributions from various household activities such as washing, toiletries, latrine usage, and bathing. This household domestic wastewater carries various nutrients and constitutes the discharged effluent. On average, about 32.5%–67.5% of domestic sewage results from everyday household activities like food preparation, drinking, personal hygiene, hot water systems, gardening and washing all of which eventually contribute to domestic wastewater that is released into the environment.

Industrial sewage wastewater, on the other hand, is comprised of runoff from industries such as paper and pulp, petrochemicals, and a range of salts, acids and chemicals, as described by Rosenwinkel et al. (2005). Industrial wastewater composition varies widely based on the contaminants and pollutants involved, necessitating various tertiary wastewater treatment methods to meet discharge regulations.

9.3   CHARACTERISTICS OF WASTEWATER EFFLUENTS

i. Physicochemical Characteristics

BOD, Dissolved oxygen (DO), pH, dissolved or suspended solids, metals and nitrogen in the forms of nitrite, nitrate, and ammonia, phosphate are the physicochemical characteristics of wastewater, as reported by Decicco (1979), Larsdotter (2006). The quality parameter of wastewater is hydrogen ion concentration (pH) and it also describes the basic or acidic properties of effluent water. Wastewater affluent in septic conditions at pH 10 indicates industrial waste and non-compatibility with biological operations. pH levels typically ranging from 6 to 9 provide a narrow range suitable for biological life, with extreme pH values having detrimental effects on biological treatment units, according to the Environmental Protection Agency (EPA) (1996), Gray (2000).

DO is another significant parameter influencing water characteristics, as it is required for aerobic microbes and other life forms to respire. The actual amount of oxygen in the solution is governed by factors like temperature, solubility, impurity concentrations such as salinity and suspended solids and atmospheric pressure, as outlined by the EPA (1996). Microorganisms use oxygen demand, in the form of BOD or chemical oxygen demand (COD), to feed on organic solids in wastewater. BOD testing requires a large number of adapter microbes and minimizes the impact of nitrifying organisms in treating toxic waste. Similarly, COD measures the oxygen equivalent of organic stuff in wastewater, including substances that undergo both chemical and biological oxidation. COD values are typically higher than BOD values due to this broader measurement range, as noted by Gray (2002).

Heavy metals are termed constant pollutants in wastewater because they accumulate in the food chain, developing health concerns and disrupting the environment. These non-degradable heavy metals can originate from groundwater infiltration, industrial discharges and residential areas. Their presence depends on various local factors such as the types of industries in the region, lifestyles, and awareness of environmental impacts from improper waste disposal, as highlighted by Hussein et al. (2005), Díaz et al. (2006).

In natural water sources, excess phosphorus may lead to eutrophication. Hence, it is crucial to control the discharge of phosphorus industrial and municipal wastewater for preventing surface water eutrophication, according to the Department of Natural Science (2006). The principal chemical test analysis which indicates the chemical nature or characteristics of the industrial wastewaters includes organic nitrogen, free ammonia, nitrates, and nitrites, inorganic and organic phosphorus. The two primary nutrients responsible for the growth of aquatic plants are phosphorus and nitrogen, as mentioned by Rein (2005).

Heavy metal levels must be determined as they can have toxic effects, and priority metal pollutants like arsenic, cadmium, chromium, and mercury need removal before considering further use of treated sludge or effluent. In industrialized nations, volatile organic compounds like toluene, benzene, trichloroethane, xylenes, trichloroethylene and dichloromethane are the general soil pollutants. Furthermore, organic compounds classified as priority pollutants encompass polycyclic aromatics and polychlorinated biphenyls (PCBs), among others, including formaldehyde, 1,3-butadiene, acetaldehyde, 1,2-dichloroethane, hexachlorobenzene (HCB), and dichloromethane.

ii. Microbiological Characteristics

Microorganisms, including bacteria, viruses, algae, fungi, helminths, and protozoa present in wastewater, have been associated with various waterborne outbreaks, as observed by Kris (2007). In addition to their potential for causing health issues, these microorganisms play a vital part in the secondary treatment of wastewater by participating in processes that leads to organic matter degradation, ultimately resulting in reduced sludge production, as noted by Ward-Paige et al. (2005). Furthermore, wastewater microbes serve as indicators of water quality due to their involvement in nutrient recycling processes, such as those related to nitrogen, phosphate, microbial pollutants and heavy metals. Detecting, isolating, and identifying different microbial pollutants in wastewater are often high-cost and time-consuming endeavors. To evaluate the relative risk of pathogen presence in wastewater, indicator organisms are utilized. Enteric bacteria, including *Escherichia coli*, fecal *Streptococci* and *Coliforms*, are frequently employed as markers for fecal contamination in water sources, as highlighted by DWAF (1993), Momba and Mfenyana (2005).

9.4 MICROBIAL COMMUNITIES IN WASTEWATER

9.4.1 Microorganisms

i. Protozoa

Within the realm of activated sludge, ciliated protozoa constitute the most abundant category of protozoa, followed by flagellated protozoa and amoebas. Notable examples of ciliated protozoa frequently encountered in wastewater treatment processes encompass species such as *Carchesium polypinum, Aspidisca costata, Opercularia coarcta, Chilodonellaun cinata, Trachelophyllum pusillum, Opercularia microdiscum, Vorticella convallaria,* and *Vorticella microstoma* as documented by Jayakumar and Natarajan (2012), Akpor et al. (2014).

ii. Viruses

Human viruses that are greatly discharged in feces are also commonly discovered in wastewater. Viruses that are inhabitant to plants and animals as well as bacteriophages exist in smaller quantities in wastewater effluent (Jayakumar & Natarajan, 2012; Akpor et al. 2014).

iii. Fungi

Zoogloea sp. and *Sphaerotilus natans* are notable fungi commonly found in sewage systems. Several filamentous fungi exist naturally in wastewater treatment systems in the form of spores or vegetative cells, and they possess the ability to metabolize organic substances. Various fungal species, including *Penicillium, Aspergillus, Absi*dia, *Fusarium, etc.,* have been utilized for carbon and nitrogen removal in wastewater treatment, as indicated by Jayakumar and Natarajan (2012), Akpor et al. (2014).

iv. Algae

Algae can also be present in wastewater due to their capacity to use nitrogen and phosphate for growth, as well as sunlight energy for photosynthesis, which can contribute to eutrophication. Algal species like *Chlamydomonas* sp., *Oscillatoria* sp., and *Euglena* sp. are known to inhabit wastewater. An alga plays a major part in the biological purification of wastewater because it can accumulate heavy metals, plant nutrients, both inorganic and organic toxic substances and pesticides. In recent years, the exploitation of microalgae in biological wastewater treatment has grown significantly as highlighted by Jayakumar and Natarajan (2012) and Akpor et al. (2014).

v. Helminths

Helminths are present in significant numbers in secondary wastewater biofilters, biological contractors and effluents. The absence of nematode activity can be a potential indicator of a toxic condition in the treatment process, particularly in developing the treatment process, as mentioned by Jayakumar and Natarajan (2012), Akpor et al. (2014).

vi. Main Purifying Bacteria

Biological purification processes owe their effectiveness to the activities of various microorganisms. In these processes, organic matter undergoes decomposition through nitrification in aerobic zones, denitrification in anaerobic zones, and the digestion of other organic compounds, such as sulfates and phosphates, often occurring in the anaerobic zone, as explained by Swiontek Brzezinska et al. (2014).

vii. Carbonate-reducing Bacteria

These anaerobic bacteria possess the ability to oxidize organic compounds, with a prominent group being methane-producing bacteria, including *Methanosarcina, Methanococcus,* non-sporulated *Methanobacterium* and sporulated *Methanobacillu*s, as detailed by Swiontek Brzezinska et al. (2014).

viii. Nitrifying and Denitrifying Bacteria

In the realm of biological purification, various bacteria play essential roles in reducing nitrogen accumulation in wastewater, eliminating nitrate through denitrification, and mitigating eutrophication in sewage water ecosystems. Notable species contributing to these processes include *Azospirillum brasilense, Achromobacter alcalinigens, Sp. psychrophilum, Spirillum lipoferum, B. licheniformis, B. azotoformans, Chromobacterium lividum, Chromobacterium violaceum, Corynebacterium nephridii, Halobacterium marismortui, Kingella denitrificans, Neisseria sicca, N. subflava, N. flavescens, Paracoccus halodenitrificans, N. mucosa, Pr. cidipropionici, Propionibacterium pentosaceum, Nirozomonas, Pseudomonas, Thiobacillus denitrificans,* and *Nitrobacter,* among others, as elucidated by Richardson et al. (2009).

ix. Dephosphatous Bacteria

Acinetobacter species, such as *A. lwoffii, A. baumannii, A. johnsonii, A. junii, A. bouvetii, A. baylyi, A. tjernbergiae, A. grimontii, A. tandoii, A. towneri, A. kyonggiensis, A. gerneri, A. pakistanensis* and *A. rudis,* are dominant in raw and sewage wastewater and play a crucial role in activated sludge systems. They facilitate the dephosphatation process through anaerobic/aerobic sequences, altering enzymatic equilibrium, also inducing phases of phosphorus accumulation. In anaerobic conditions, these bacteria release phosphorus, which is then reabsorbed as oxygen levels rise. The bacterium can use polyphosphate assimilation as either phosphorus or an energy reserve, as detailed by Yazdani et al. (2009), Al Atrouni et al. (2016).

x. Sulfate-reducing Bacteria

Sulfate-reducing bacteria, including *Desulfobacterium autotrophicum, Desulfobulbus propionicus,* and *Desulfovibrio desulfiricans,* are anaerobic heterotrophic and autotrophic bacteria that contribute significantly to organic matter biocorrosion, sulfate reduction and mineralization in wastewater treatment processes. They help in the breakdown of sediments in wastewater and also conversion of sulfate to sulfide, as explained by Kumar et al. (2011), Tingting and Dittrich (2016).

9.5 ROLES AND DYNAMICS OF MICROORGANISMS IN WASTEWATER TREATMENT SYSTEMS

Treatment systems microbial populations in wastewater treatment systems encompass protozoa, bacteria, fungi, viruses, helminths, and algae. However, the occurrence of many of these organisms in wastewater can lead to the spread of infectious diseases.

i. Bacteria

Bacteria play a crucial role in wastewater treatment systems by facilitating the conversion of complex organic matter into simpler compounds. These microorganisms typically range in size from 0.2 to 2.0 μm in diameter and are primarily responsible for the efficacy of wastewater treatment, especially in septic tanks, as noted by Stevik et al. (2004). While not all bacteria are harmful, some can cause waterborne diseases in animals and humans. Examples of such diseases include dysentery, cholera, salmonellosis, gastroenteritis and typhoid fever, as highlighted by Jenkins et al. (2004). In wastewater treatment systems, bacteria are often found within flocs, as seen in activated sludge systems, where some play essential roles in biological treatment processes. However, certain filamentous bacteria can pose challenges by interfering with settling and causing foaming, as pointed out by Eikelboom (2000). Waterborne gastroenteritis of unknown origin is frequently reported, with certain strains of bacteria, including *E. coli and Pseudomonas* sp., posing a potential risk to newborns and being associated with epidemics of digestive disease, as indicated by Metcalf and Eddy (2003) (Table 9.1).

Bacteria hold paramount importance in wastewater treatment systems, with many of them being facultative, capable of thriving in the absence or presence of oxygen, as noted by Absar et al. (2005). While both autotrophic and heterotrophic bacteria are present in these systems, heterotrophic bacteria are predominant. The carbonaceous organic matter in wastewater effluent provides energy to heterotrophic bacteria, using it for cell synthesis and energy release through the breakdown of water and organic matter. Several significant genera of bacteria found in wastewater treatment systems include *Alcaligenes, Achromobacter, Citromonas, Arthrobacter, Pseudomonas, Flavobacterium, Acinetobacter* and *Zoogloea*, as reported by Water Environment Association (1987), Oehmen et al. (2007), EPA (1996).

In wastewater treatment, bacteria are accountable for stabilizing influent waste, often forming floc particles that aid in waste breakdown. These floc particles also serve as sites for the absorption and breakdown of waste materials. Filamentous bacteria, organized into filaments or trichomes, provide structural support for the floc particles, enabling them to withstand the shearing forces and increase in size during the treatment process. However, excessive or lengthy filamentous bacteria can lead to problems related to solid/liquid separation or settleability, as explained by Paillard et al. (2005), Gray, (2002).

Furthermore, bacteria are a common source of microbial pollution in wastewater. Total and fecal coliforms testing can reveal the presence of harmful microorganisms, as outlined by EPA (1996), APHA (2012). Traditionally, the detection of fecal coliforms is widely accepted as a reliable indicator of fecal contamination, with *E. coli* being considered a dependable indicator for fecal pollution from both human and animal sources, as it is known to have a limited lifespan outside of the fecal environment, according to Odonkor and Ampofo (2013). Tests for total and fecal coliforms can be conducted using either traditional methods or enzymatic techniques. Membrane filtration techniques and multiple-tube fermentation are the conventional approaches, with the former being suitable for moderately to highly contaminated waters and the latter for waters with low to very low contamination levels.

TABLE 9.1

Disease Caused by Microorganisms Present in Sewage Wastewater and their Causative Agents

Microorganisms	Diseases
Bacteria *Clostridium perfringens, Campylobacter* spp., *E. coli, Pseudomonas aeruginosa, Helicobacter* spp., *Enterococci* spp., *Legionella* spp., *Klebsiella pneumonia, Clostridium tetani, Leptospira*spp., *Salmonella typhi, Proteus* spp., *Salmonella paratyphi, Salmonella typhimurium, Staphylococcus aureus, Shigelladysenteriae, Vibrio cholera, Streptococcus* spp.,	Typhoid, bacillary dysentery, paratyphoid, cholera, epidemic hepatitis, meningitis, gastroenteritis, colitis, vomiting, erythema, enteritis, inflammation of the urinary tract, diarrhea (Enterotoxin), pus accumulation, tetanus, respiratory tract inflammation, local infection, otitis, pneumonia, and endocarditis
Fungi *A. fumigates, Aspergillus* spp. *Trichophyton* spp., *A. niger,*	Mycosis of the nails, broncho-pulmonary mycosis, granuloma, otitis, and mycosis of the skin
Nematodes *Anclyostomaduodenale* *Ascarislumbricoides* *Toxocaracanis*	Infect lungs, the small intestine, internal bodies, and intestine
Protozoa *Giardia lamblia (Lambliaintestinalis) Entamoeba histolytica* *Sarcocystis* spp. *Cryptosporidium* spp. *Toxoplasma gondii*	Infect liver, giardiasis, biliary gall bladder, internal organs (liver, brain, heart), cryptosporidiosis, and intestine
Tapeworms *Echinococcusgranulosus* *Taeniasolium* *Taeniasaginata*	Infect liver, intestine, and lungs
Virus *Coxsackie, Polio, Hepatite A, B and C, Echo, Coronavirus, Adenovirus, Enterovirus*	Fever, poliomyelitis, meningitis, respiratory diseases, hepatitis, myocarditis, eye inflammation, cold, and encephalitis
Yeast *Cryptococcus neoformans* *Candida crusei* *Candida albicans*	Disorders of the mucous membranes (mouth), meningoencephalitis, granuloma, and lungs

ii. Protozoa

Protozoa, tiny single-celled organisms, are a common presence in wastewater treatment systems, contributing significantly to the treatment process. They serve various crucial functions, including the clarification of secondary effluent by removing bacteria, acting as bioindicators for sludge health and flocculating suspended materials. Protozoa found in wastewater treatment systems are generally proficient for movement at some stage in their life cycle, as explained by Amaral et al. (2004).

These organisms are about ten times larger than bacteria, comprising unicellular entities with membrane-enclosed organelles. Notably, protozoa play a role in preying upon pathogenic bacteria, providing a distinct advantage in wastewater treatment, as highlighted by Mara and Horan (2003). Protozoa can be categorized into five groups based on their mode of locomotion: (i) amoeboid forms, (ii) free-swimming ciliates, (iii) flagellates, (iv) stalked and sessile ciliates, and (v) crawling ciliates, as outlined by Caccio et al. (2003) (Table 9.1).

Protozoa serve as valuable biological indicators in wastewater treatment systems. While some protozoa can survive for up to 12 hours in the absence of oxygen, they are generally regarded as obligate aerobes. Consequently, they excel at indicating an aerobic environment and the presence of potential toxicity. Protozoa often exhibit higher sensitivity to toxicity compared to bacteria. The absence or immobility of protozoa can signify potential toxicity within a treatment system. The presence of well-evolved protozoa in significant numbers within the biological mass is considered a hallmark of a well-operated and stable wastewater treatment system, as elucidated by Fried and Lemmer (2003).

Protozoa are classified into different groups based on their means of locomotion. Three of these groups possess hair-like structures or cilia that beat in coordination, creating water currents for capturing bacteria and facilitating movement. In aeration tanks of biological systems, these populations engage in continuous competition for food resources. The growth of decomposers, primarily heterotrophic bacteria, depends on the quality and quantity of dissolved organic matter, while the growth of predators relies on the availability of prey. Relationships of competition and predation give rise to population oscillations and successions until dynamic stability is achieved. This dynamic is closely tied to plant management choices designed to ensure optimal efficiency. Some ciliates are predators of other ciliates or exhibit omnivorous behavior, consuming a variety of organisms, including flagellates, dispersed bacteria and small ciliates. Bacterivorous ciliates use ciliary currents to direct suspended bacteria toward their oral region. In activated sludge, ciliated protozoa are the most prevalent, although flagellated protozoa and amoebas may also be present. Commonly observed species of ciliated protozoa in wastewater treatment processes include *Carchesium polypinum*, *Aspidisca costata*, *Opercularia coarcta*, *Chilodonella uncinata*, *Trachelophyllum pusillum*, *O. microdiscum*, *V. microstoma*, and *Vorticella convallaria* as noted by Caccio et al. (2003), Amaral et al. (2004).

Protozoa in wastewater treatment systems can be further classified based on their locomotion characteristics. Free-swimming ciliates like *Paramecium* sp. and *Litonotus* sp. possess cilia covering their entire body surface and are typically found suspended in the bulk solution. In contrast, crawling ciliates, such as *Aspidisca* sp., and *Euplotes* sp. have cilia only on their belly or ventral surface, where the mouth opening is located. These crawling ciliates are commonly associated with floc particles.

Stalked ciliates, like *Carchesium* sp., and *Vorticella* sp. have cilia concentrated surround the aperture of the mouth and are connected to floc particles. They have an expanded anterior portion or narrow posterior part, with cilia and stalk action creating a water vortex that draws dispersed bacteria into the mouth opening, as described by Gerardi (2007). Additionally, there are two types of amoebas in wastewater treatment systems: naked amoebas, including *Thecamoeba* sp. and *Actinophyrs* sp., and testate amoebas, such as *Cyclopyxis* sp., which have a protective covering composed of calcified material, according to Richard et al. (2003). Flagellated protozoa, oval-shaped organisms with whip-like flagella, move through wastewater treatment systems by propelling themselves with flagellar motion in a corkscrew pattern.

iii. Viruses

Viruses are also present in wastewater, primarily human viruses that are excreted in feces in significant quantities. While animal and plant viruses naturally exist in smaller quantities in wastewater, bacterial viruses may also be present, as reported by Okoh et al. (2007). These viruses are responsible for various water-related infections in humans, including respiratory and gastrointestinal infections, meningitis and conjunctivitis. Enteric viruses are the causative agents of a majority of waterborne diseases with unidentified sources, as detailed by EPA (1996) (Table 9.1). Viruses in wastewater are notoriously persistent and can remain viable sources of infection for months after entering the host, according to Santamaria and Toranzos (2003).

iv. Fungi

Fungi are the predominant microorganisms commonly encountered within wastewater treatment systems. Unlike bacteria, fungi are multicellular organisms that play an essential role in activated sludge. They are proficient in metabolizing organic compounds and, under specific environmental conditions in a mixed culture, can outcompete bacteria. Some fungi even possess the capability to oxidize ammonia to nitrite or nitrate. The two most prevalent fungal species in sewage are *Zoogloea* sp. and *Sphaerotilus natans* as indicated by LeChevallier and Au (2004) (Table 9.1).

In wastewater treatment systems, several filamentous fungi naturally exist as spores or vegetative cells, and they can also metabolize organic substances. Certain fungal species, including *Penicillium*, *Aspergillus*, *Absidia*, *Fusarium*, and others, have been identified for their role in removing carbon and nutrient sources from wastewater, as highlighted by Akpor et al. (2013). In systems with low pH levels where bacterial growth is inhibited, fungi primarily function in the decomposition of organic matter. Additionally, some fungi employ their fungal hyphae to trap and adsorb suspended solids, fulfilling their energy and nutrient requirements. Certain filamentous fungi have been reported to secrete enzymes that aid in substrate degradation during wastewater treatment, as documented by Molla et al. (2004).

v. Algae

Algae are present in wastewater due to their ability to harness solar energy for photosynthesis, along with their utilization of phosphorus and nitrogen for growth, which can contribute to eutrophication. Various types of algae, such as *Chlamydomonas* sp., *Oscillatoria* sp., and *Euglena* sp., can be found in wastewater. Algae hold significant importance in the biological purification of wastewater because they can accumulate plant nutrients, heavy metals, pesticides, and both organic and inorganic toxic substances. The utilization of microalgae in biological wastewater treatment has gained increasing significance over the years, as noted by Lloyd and Frederick (2000). High-rate algal ponds, characterized by their shallow depth and mechanical aeration with paddle wheels, are capable of treating 80% of nitrogen and phosphorus and 90% of BOD in wastewater (Table 9.1). These ponds offer cost advantages in terms of construction, energy, and land requirements compared to facultative ponds and constructed wetlands, as illustrated in Table 9.1.

vi. Helminths

Nematodes are aquatic organisms that inhabit fresh, brackish, and salt-water environments, as well as damp or moist soils globally. Freshwater nematodes are found in sand filters and aerobic treatment facilities, and in, biofilters, biological contractors and secondary wastewater effluents, these nematodes are present in significant numbers. They exist below the water table and rely on dissolved oxygen in freshwater. Nematodes contribute to the ecosystem as they serve as food for small invertebrates, as elucidated by Metcalf and Eddy (2003). These nematodes move in a whip-like fashion when in their free-living mode and secrete a sticky substance to anchor themselves to a substrate, allowing them to feed without disruption from turbulence or currents. A lack of nematode activity can serve as a bio-indicator of a toxic condition within a treatment system, as shown in Table 9.1.

Within the helminth group, there are many species of parasitic worms that can cause severe illnesses. These helminths fall into three categories: roundworms, annelids, and flatworms. Flatworms are further divided into two subgroups: tapeworms, which have segmented bodies, and flukes, which possess flat, unsegmented bodies. Most helminths reproduce through eggs, although the size and shape of the eggs can vary. Notably, many helminth eggs in wastewater are not inherently infectious. To become infectious, these eggs must be viable and undergo larval development, a process influenced by moisture and temperature levels. Helminth eggs can remain viable for 1–2 months in crops, even longer in soil, freshwater and sewage, and potentially years in sludge and feces Nelson et al. (2004), Sanguinetti et al. (2005), Kone et al. (2007). Inactivation of helminth eggs occurs at elevated temperatures (above 40°C) and reduced moisture levels (below 5%), conditions not typically attained during wastewater treatment. Traditional water treatment processes like sedimentation, filtration, or coagulation-flocculation can effectively remove helminth eggs from the water, as supported by Jimenez (2007).

9.6 BIOLOGICAL PROCESSES FOR WASTEWATER TREATMENT

The primary objective of wastewater treatment is to prevent water source pollution and safeguard public health by curbing the spread of diseases. Various wastewater treatment systems, including onsite and offsite systems, are employed to achieve this goal. In this section, we will focus on describing offsite wastewater treatment systems, such as constructed wetlands, membrane bioreactors, activated sludge, trickling filters and stabilization ponds, as outlined by Stottmeister et al. (2003). Offsite treatment systems rely on microorganisms to convert organic contaminants into less toxic compounds, ultimately mineralizing them into water, inorganic salts and carbon dioxide within specific bioreactors. However, it's important to note that these processes come with certain drawbacks, including increased labor requirements and the need to pump the water to be treated WHO (2006). Additionally, these methods may not be well-suited for treating high concentrations of highly toxic or poorly biodegradable pollutants, as depicted in Figure 9.1.

i. Activated sludge

The activated sludge process is a method for removing organic matter from wastewater, utilizing a high concentration of microorganisms, predominantly protozoa, fungi and bacteria. These microorganisms form loose clumped masses of fine particles suspended in the wastewater, with continuous stirring, as described by Templeton and Butler (2011). Activated sludge, essentially sewage rich in vigorous microorganisms, plays a pivotal role in organic matter breakdown. Among various wastewater treatment systems, activated sludge stands out as the most versatile and effective method. In an activated sludge system, microorganisms in the aeration tank of the wastewater break down the organic matter. These microorganisms consist of 10%–30% inorganic matter and 70%–90% organic matter, and their

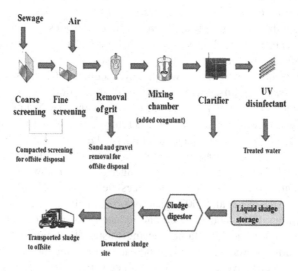

FIGURE 9.1   Schematic diagram of sewage wastewater treatment process.

characteristics vary based on the chemical conditions and specific organism types in the biological mass. After the mixed liquor is discharged from the tank, suspended solids from the treated wastewater are separated by gravity in a clarifier (also referred to as a settling or sedimentation tank). The concentrated biological solids are then recycled back to the aeration tank to maintain a concentrated microorganism population. To manage the excess biological solids produced, a method for their dispersion must be provided, as microorganisms are continually reproducing in the system. Wasted solids from the aeration tank have a lower concentration compared to those from the clarifier, necessitating the handling of a larger sludge volume. The production of solids can be maximized or minimized depending on the process's design and operation, as stated in Sci-Technology Encyclopedia (2007).

A typical activated sludge process includes an aeration tank, a mechanism for oxygen transfer to microorganisms, a means of stirring the fluid mixture in the aeration tank, a system for separating microorganisms from treated water, and a microorganism recycling system. Microorganisms' activity aids in the oxidation of organic matter in sewage into carbon dioxide and water. During the initial treatment stages, large floating materials are screened out, and wastewater is then allowed to pass through a settling chamber for further treatment, removing sand and other materials. In the system's aeration tank, air is introduced into the effluent for primary treatment. Activated sludge has the capacity to remove 75%–90% of the BOD from sewage, as indicated by Schmidt et al. (2002), Tortora et al. (2010) (Figure 9.2).

In the activated sludge process, the carbonaceous organic matter present in wastewater serves as the energy source for producing new cells. Microorganisms convert this carbonaceous organic matter into cell tissue and oxidize end products, including carbon dioxide, sulfate, nitrate, and phosphate. Activated sludge systems are widely used to reduce the concentration of particulate, colloidal and dissolved, organic pollutants in wastewater. Design parameters are primarily based on the solid retention time (SRT) or the food to microorganism ratio (F:M), along with the hydraulic detention time, and these parameters are influenced by the length of time sludge is retained in the system Gray (2002).

Activated sludge processes can be tailored for various objectives, such as degrading carbonaceous BOD, nitrification (oxidizing ammonia to

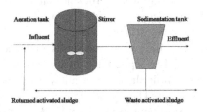

FIGURE 9.2   Illustration of activated sludge process.

nitrate), and denitrification (reducing nitrate to nitrogen gas). Modifications to some systems include phosphorus removal and biological nitrification. These processes create an ideal environment for the multitude of microorganisms present in activated sludge systems through constant aeration, agitation, and recirculation, while restraining the growth of larger microorganisms. Activated sludge systems typically contain bacteria, fungi, protozoa, rotifers, and nematodes, though not all may coexist in a single system. Bacteria, known to consume organic matter in wastewater, are generally the most significant organisms present. Algae, due to their light requirements, are rarely found in mixed liquor. The collective metabolism of all microorganisms in the activated sludge governs the overall reactions taking place in the system.

The metabolic process involves separate yet simultaneous synthesis and respiration reactions. Synthesis entails using a portion of the waste matter (food) to create new cells (protoplasm), while respiration releases energy by converting food material into lower-energy compounds, typically water, carbon dioxide and potentially various oxidized nitrogen products. The exact nature of the resulting products relies on factors like process design, reaction time, temperature, and system loading, as Sci-Technology Encyclopedia (2007).

ii. **Trickling filter**

The trickling filter is a commonly employed method for secondary wastewater treatment. This process involves a filter bed constructed from highly porous media, such as plastic materials or gravel, which forms a layer of microorganisms, leading to the development of a slime layer. Within the trickling filter system, microorganisms adhere to the media and establish a biofilm. As wastewater flows through this media, the microorganisms within the biofilm consume and eliminate contaminants, as described by Tchobanoglus et al. (2003).

In a trickling filter system, sewage is distributed over the permeable media, which can consist of materials like molded plastic, rocks, ceramics and gravel. The media must possess a size that permits the passage of air to the bottom while maximizing the available surface area for microbial activities. As air circulates through the media, an aerobic microorganism biofilm forms and grows on its surface. In the slime layer aerobic bacteria metabolize organic materials to produce water and carbon dioxide. It's important to note that this treatment system typically removes approximately 80%–85% of the BOD, rendering it less efficient than activated sludge systems. Nonetheless, trickling filters are straightforward to operate and are not susceptible to issues related to toxic sewage.

A complete trickling filter setup comprises a clarifier, a septic tank, and an application system. The septic tank serves to remove solid components from the wastewater, while the clarifier allows biological materials to settle out of the wastewater. The application system aids in distributing the treated wastewater to the appropriate location. Prior to entering a trickling filter, wastewater must undergo pretreatment to eliminate solid and greasy materials. This pretreatment step is crucial to prevent these substances from covering the thin layer of microorganisms and potentially harming them.

Trickling filters can be categorized as a low rate or high rate based on organic loading or hydraulic. Low-rate filters are employed for straightforward treatment, consistently producing a certain effluent quality. These low-rate trickling systems are designed to remove 80%–85% of the applied BOD. In contrast, high-rate filters are characterized by higher hydraulic and organic loadings compared to low-rate filters. High-rate filters involve a recirculation process, where filter effluent is returned and reapplied to the filter. This recycling of wastewater increases the contact between microorganisms and waste, enhancing the efficiency of effluent treatment, as discussed by Van Haandel and Van Der Lubbe (2007).

iii. **Membrane bioreactor**

A membrane bioreactor (MBR) is an innovative wastewater treatment system that combines the biological degradation process of activated sludge with direct solid-liquid separation using micro or ultrafiltration membrane technology. This system achieves complete physical retention of bacterial flocs and all suspended solids within the bioreactor. MBRs offer several advantages over other treatment methods, including effective disinfection capabilities, superior effluent quality, reduced sludge production, and higher volumetric loading capacity. They represent a biological wastewater treatment process employing membranes to replace the gravitational settling phase of the conventional activated sludge process for solid-liquid separation of sludge suspension. MBRs are employed for the treatment of biologically active wastewater from industrial or municipal sources. Two configurations of MBRs exist external/sidestream and internal/submerged. In the submerged configuration, the membranes are immersed within the biological reactor, while in the external/sidestream configuration, the membranes constitute a separate unit process that involves intermediate pumping steps, as explained by Singhania et al. (2012), Lofrano et al. (2013).

The membranes in a membrane bioreactor system are typically composed of inorganic materials or polymers, featuring numerous tiny pores that require microscopic observation for detection. These small pore sizes selectively allow the passage of very minute particles and water molecules while retaining larger contaminants, as discussed by Tom et al. (2000). Common membrane configurations employed in MBRs include flat sheets, tubular membranes, and hollow fiber. Flat sheet membranes and hollow fiber are immersed in the water, while tubular membranes are typically positioned outside the bioreactor, as suggested by Cornel and Krause (2008). In MBRs, two primary challenges are sludging and fouling. Sludging refers to the accumulation of soft, wet mud, thick, often stemming from industrial waste, which can impact membrane surface area. Preventive measures include ensuring adequate water flow within the medium, as recommended by Radjenovic et al. (2007). On the other hand, fouling occurs due to the buildup of particles on the membrane surface, resulting in increased resistance during filtration. Fouling control strategies encompass the installation of aerators beneath the membrane, as outlined by Sutton (2006). Periodic relaxation and backflushing operations are also performed to remove fouling layers from the membrane surface in membrane filtration.

MBR technology competes with other biological wastewater treatment systems, notably the conventional activated sludge process. While conventional processes are effective in meeting standard discharge requirements and are cost-efficient, they may struggle to meet stringent treatment standards for sensitive environments. Additionally, conventional methods may not be economically viable for wastewater reuse unless post-treatment with microfiltration or ultrafiltration membranes is applied, as highlighted by UNEP (2012).

iv. **Stabilization pond**

One of the most significant natural approaches to wastewater treatment is the use of wastewater stabilization ponds. These ponds are typically shallow man-made reservoirs that may consist of one or several series of facultative, anaerobic, and maturation ponds. The treatment process in these ponds follows a sequential pattern, starting with the anaerobic pond, which is primarily designed for the removal of suspended solids and some organic matter. The next stage, referred to as the facultative pond, further eliminates organic matter through the activities of heterotrophic bacteria and algae, including species such as *Rhodococcus* sp., *Pseudomonas* sp., and *Arthrobacter* sp., as noted by Diep and Cuc (2013). Finally, the maturation pond, the last stage, focuses on the removal of pathogens and nutrients. This system is renowned for its cost-effective approach to wastewater treatment, relying on natural disinfection mechanisms. Stabilization ponds are particularly well-suited for subtropical and tropical regions, thanks to the higher temperatures and abundant sunlight that enhance treatment processes, as emphasized by Pena-Varon (2002).

Anaerobic ponds represent the initial stage in the series of stabilization ponds. These ponds are relatively small, typically 2–5 m deep, and receive wastewater rich in organic matter. The high organic content fosters strict anaerobic conditions, meaning that dissolved oxygen is absent within the pond. Anaerobic ponds can be likened to open septic tanks in terms of function. A well-designed anaerobic pond can achieve an impressive 60% removal of organic matter, especially at a temperature of 20°C, as outlined by Shilton and Harrison, (2003). These ponds are highly efficient, require minimal space, and produce nutrient-rich sludge suitable for various applications. In warmer conditions, the removal of BOD can reach as high as 60%–85% within a short timeframe. Facultative ponds come in two primary types: primary facultative ponds, which receive untreated wastewater, and secondary facultative ponds, which receive settled water from the initial anaerobic pond. These ponds are designed with a low organic surface load to allow the development of algal populations. Algae play a crucial role in generating the necessary oxygen to remove soluble organic matter. As a result, the water in facultative ponds often takes on a dark green hue, although it can occasionally appear red or pink due to the presence of purple sulfide-oxidizing photosynthetic activity. Changes in color within facultative ponds serve as indicators of ongoing removal processes. In some cases, the color alteration may be attributed to the presence of red algae, such as species of *Rhodophyta*. The wind's velocity is a critical factor as it promotes the mixing of pond water, ensuring uniform distribution of dissolved oxygen, organic matter, algae, and bacteria, as discussed by Mara et al. (2007).

v. **Constructed wetland**

Constructed wetlands are ones into which wetland vegetation is purposely placed to enhance pollutant removal from stormwater runoff. Constructed wetlands are ditched basins with irregular perimeters and undulating bottom contours. Through a forebay, stormwater enters a constructed wetland where the coarse organic materials and larger solids settle out. From the forebay, the stormwater discharges and passes through emergent vegetation that filters soluble nutrients and organic materials. Some dissolved nutrients are removed through the means of vegetation. To reduce peak, stormwater flows the constructed wetlands are designed by Tchobanoglus et al. (2003).

The use of constructed wetlands is in two ways. First, constructed wetlands are used principally to maximize the removal of pollutants from stormwater runoff and it also helps in controlling stormwater flows. With increased pollutant removal capabilities, it may also be used primarily to control stormwater flows. If the same area were twisted into a rectangular stormwater basin, the secondary benefits of constructed wetlands comprise increased wildlife habitats, preservation and restoration of the natural balance between surface waters and ground waters and elevated property values.

This constructed wetland system takes advantage of processes that exist in natural wetlands. Subsurface flow and free-water surface systems are the two types of treatment systems in constructed wetlands. In the free-water surface system, from wastewater, the pollutants are detached by the decomposing microorganisms (mostly bacteria and fungi) living on aquatic plants and soil surfaces. The microorganism attached to the aquatic plants below the water level utilizes oxygen during decomposition. The aquatic plants play a vital role in the uptake of nutrients, such as nitrogen, phosphorus, and other compounds from the wastewater apart from helping in the decomposition process alone. Some of the nitrogen and phosphorus are released back into the water when the plants die and decompose. Numerous constructed wetland systems have been prepared and run to improve water quality, in order to provide high-quality wetland habitat for migratory birds and also used for wastewater treatment, reuse, and disposal systems.

In the subsurface flow system, it is planned to create subsurface flow by keeping the water that is treated below the surface through a permeable medium. This assists in evading the development of bad odors and others. Usually, the hydraulics of the system are affected because the media are typically soil, sand, gravel, or crushed rock. In free-water surface system, it is designed to replicate through a shallow depth of the natural wetlands with water flowing. The two types of wetlands treatment systems are designed to limit leaching in basins or channels with a constructed or natural subsurface barrier (Akpor et al. 2014).

vi. Rotating biological contactors

Rotating biological contactors (RBCs) are made up of vertically arranged plastic media that are placed on a horizontally rotating shaft. The plastic media typically have a diameter ranging from 2 to 4 mm and a thickness of up to 10 mm. The shaft rotates slowly, usually at 1–1.5 revolutions per minute (rpm). This rotational speed is essential to provide hydraulic shear for sloughing and to maintain turbulence that keeps solids in suspension. Approximately 40% of the media is submerged, and the biomass-coated media is alternately exposed to ambient oxygen and wastewater as the shaft rotates. The high surface area of the media allows for the development of a stable biomass population that continuously undergoes surplus growth. This excess growth is automatically shed and subsequently removed in a downstream clarifier. Depending on the strength of the wastewater and the spinning speed of the disk, the thickness of the biofilm can range from 2 to 4 mm. While RBC systems are well-suited for municipal wastewater treatment, they are relatively new compared to other technologies due to their ability to quickly recover from upset conditions. These systems have found applications in various petroleum facilities, as noted by Schultz (2005). RBC systems are highly adaptable and can be easily expanded. They are also effective at containing volatile organic content. In terms of power requirements, RBCs have relatively low demands and can even be powered by compressed air for aeration. They have straightforward operating procedures but do require a reasonably trained workforce. However, it's important to note that RBCs can be intensive to install and are sensitive to temperature fluctuations.

vii. Oxidation ditches

Oxidation ditch is a customized activated sludge biological treatment method that eliminates biodegradable organics by utilizing a sludge age of 12–20 days and a hydraulic retention period of 24–48 hours. Oxidation ditches are typically designed as complete mix systems, although they can be modified as needed. The configuration of an oxidation ditch can be single or multichannel, with preliminary treatment processes like bar screens and grit removal typically preceding the oxidation ditch. Depending on effluent requirements, primary settling may sometimes be carried out before the oxidation ditch, and tertiary filters may be required after clarification. Disinfection is necessary prior to final discharge, and reaeration is often employed. Within the ditch, circulation, oxygen transfer, and aeration are facilitated by horizontally or vertically mounted aerators. Aeration occurs through the flow of wastewater into the oxidation ditch and is mixed with return sludge from a secondary clarifier. This mixing process introduces oxygen into the mixed liquor, promoting the growth of microorganisms. The motive velocity helps ensure contact between microorganisms and influents. While aeration during mixing increases dissolved oxygen concentration, it decreases as biomass consumes oxygen. Solids remain in suspension during circulation, as noted by US-EPA (2000) (Figure 9.3).

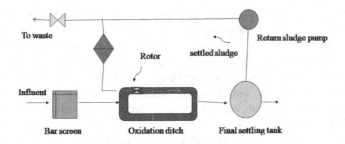

FIGURE 9.3 Schematic diagram of oxidation ditches.

9.7 APPLICATIONS OF TREATED WASTEWATER

9.7.1 USE OF TREATED WASTEWATER IN IRRIGATION

In regions grappling with water scarcity, factors such as prolonged droughts, population expansion, limited water resources, and the worsening condition of freshwater due to extensive waste discharge have driven the adoption of wastewater for irrigation purposes. Recycled wastewater, rich in nutrients, organic matter, and fertilizers, stands out as a premium resource for augmenting plant growth and enhancing crop yields (Baduru & Sai, 2015; Djadouni & Madani, 2016). The careful management of wastewater yields several advantageous outcomes, including the recycling of essential nutrients, the irrigation of crops, the reduction of harmful contaminants in both soil and plants, cost savings in artificial fertilizer usage, and the mitigation of pathogenic microorganisms in the surrounding environment. However, the neglect of wastewater management can result in two significant water quality issues: chemical contamination and microbial pollution. This contamination affects the soil adversely by introducing both desirable and undesirable chemical constituents, along with salinity, which can taint crops and compromise soil health. Additionally, the presence of microbial pathogens in wastewater can lead to waterborne diseases, creating both environmental and health crises. These risks underscore the importance of properly managing wastewater to safeguard the ecosystem and public well-being, thereby avoiding the consequences associated with the direct or indirect use of untreated wastewater for irrigation.

9.7.2 WASTEWATER IRRIGATION: DYNAMICS OF MICROORGANISMS IN SOIL

Wastewater irrigation exerts a profound impact on soil characteristics and quality, encompassing aspects like the existence of heavy metals (such as copper, cadmium, selenium, lead, sodium, and zinc), the rate of organic matter, and the diversity of nutrients. Consequently, these factors influence various microbiological parameters, including microbial activities, enzyme activities, and microbial biomass, as well as the dynamics of pathogenic and indicator microorganisms within the soil (Wang et al. 2010; Agnieszka & Magdalena, 2016). Within the activated sludge process, a biologically diverse community of microorganisms, comprising aerobic and anaerobic species such as archaea, bacteria, protists, and fungi, plays a pivotal role.

This community exhibits the capacity to break down a wide array of organic compounds, including toluene, benzopyrene, and petroleum products the neutralization of chemical pollutants like xenobiotics and toxicants (Shumaila et al. 2013). The organisms harnessed in the activated sludge process encompass specific bacterial strains that rely on oxygen for their development. They operate synergistically to efficiently corrupt sludge, aid in bioremediation, also purify contaminated water, even involving heavy metals like sulfur, iron, ferrous sulfate and ammonium ferrous sulfate (Gopinath et al. 2015).

Five foremost kinds of microorganisms are typically discovered in the aeration basin of the activated sludge process:

- Bacteria, which are aerobic and responsible for the decomposition of organic nutrients.
- Protozoa, which eliminate and digest discrete bacteria and suspended particles.
- Metazoa, serving to control longer-aged systems, including lagoons.
- Filamentous bacteria, which may cause poor settling and turbid effluent (known as bulking sludge).
- Fungi and algae, which are present during pH fluctuations and in older sludge (Khan, 2011).
- Notably, *Bacillus* and *Pseudomonas* species are frequently employed microorganisms in the activated sludge process (Khan, 2011).

Moreover, microorganisms like *Trichoderma harzianum*, *B. pumilus* QST 2808, *Bacillus subtilis* QST 713, *Pseudomonas chlororaphis* MA 342, and *B. megaterium* ATCC 14581 serve as bacteria with fungitoxic and fungistatic properties (Abriouel et al. 2011; Karthiga Devi & Natarajan, 2015; Khaled & Balkhair, 2016). These microorganisms also produce bioactive substances that bolster plant defenses, facilitate cellulose decomposition, and assist in breaking down other organic waste components (Shakir et al. 2017). In water environments like storage basins, a variety of ubiquitous microorganisms exist, capable of causing non-enteric infections. These microorganisms include species of bacteria from genera such as *Pseudomonas*, *Campylobacter*, *Aeromonas*, *Mycobacterium*, *Legionella*, *Streptomyces*, *Bacillus*, and *Leptospira*. They play vital roles in enhancing nutrient uptake by plants, increasing the bioavailability of nutrients in the soil, inducing and accelerating organic matter degradation, and synthesis of a mixture of bioactive molecules and biocontrol agents with antifungal and antibacterial activities (Benoit et al. 2012; Deepak et al. 2014).

9.7.3 WASTEWATER AND PLANTS

Furthermore, wastewater irrigation is commonly employed for specific vegetable crops like mint, lettuce, potatoes, parsley, squash, cucumbers, watercress, radishes, and pumpkins. Wastewater contains encompassing organic matter, elevated levels of nutrients, micronutrients, inorganic compounds, phosphorus, nitrogen, and various microorganisms essential for agricultural crop growth and yield enhancement

(Chun et al. 2015; Liang et al. 2016). These nutrients, dissolved minerals, nitrogen, phosphorus, potassium, sodium, calcium, iron, as well as compounds such as sugars, proteins and fats, constitute the primary source of soil fertilization via wastewater irrigation. Soil microorganisms utilize these substances as both a source of nourishment and energy to synthesize cellular components and sustain vital life processes (Wang et al. 2010). Wastewater from treated sources is deployed for the irrigation of crops not intended for direct human consumption, a practice referred to as restricted irrigation. This includes industrial crops (e.g., sisal, cotton, sunflower), crops that undergo processing before consumption (e.g., oats, barley, wheat), fodder crops, pastures, and fruit trees. In contrast, unrestricted irrigation pertains to the cultivation of all crops destined for direct human consumption, including those consumed raw (e.g., salads, lettuce, cucumbers), and extends to the irrigation of sports fields and public parks (Djadouni, 2019).

9.8   SUMMARY AND CONCLUSION

Microbes are vital in wastewater management and treatment owing to their extraordinary capability to break down organic contaminants and enhance water quality. Microbial communities, which include bacteria, algae, and fungi can break down organic debris, remove nutrients such as nitrogen and phosphorus, and even diminish the amount of diseases in wastewater. These bacteria convert contaminants into harmless byproducts through processes such as biological nutrient removal, anaerobic digestion, and activated sludge treatment, ensuring the safe discharge of treated water into the environment or its recycle for various uses. Using microorganisms to treat wastewater not only upgrades sustainable water resource management but also minimizes the environmental footprint of traditional treatment technologies, making it a prospective path in order to address the current global issue of water pollution and scarcity.

The primary objective of wastewater treatment is to prevent the transmission of diseases by safeguarding water sources against pollution. Wastewater treatment serves as a pivotal strategy in the management of water quality. Over the years, due to certain drawbacks associated with chemical treatment methods, biological treatment approaches have gained prominence to mitigate adverse conditions in natural water resources. The management of biological nitrogen and phosphorus in wastewater treatment systems has received extensive attention. Numerous studies have revolved around nutrient removal, primarily focusing on the presence of bacteria and their well-documented roles in eliminating phosphorus and nitrogen.

Protozoa's contribution to nutrient removal in wastewater treatment has also evolved considerably in recent times. While protozoa were traditionally recognized for their effectiveness in the purification process by consuming bacteria and thereby reducing their population, their role has expanded to encompass nutrient mineralization. The proliferation of a large number of microorganisms actively involved in nutrient removal has been identified as a contributor to water source eutrophication. Consequently, there is an urgent need for increased research and monitoring efforts, specifically tailored to nutrient-related aspects. This will facilitate the achievement of uncontaminated wastewater discharge into natural water bodies, aligning with the effluent standards and regulations established by governing bodies. Enhanced clarity is essential for comprehending and elucidating the observed microbial life in wastewater treatment systems, particularly in constructed wetlands, considering factors like migratory birds.

Fungal activity has led to improvements in the settleability, dewaterability, and degradability of wastewater sludge, contributing significantly to sludge management strategies. Advancements in molecular techniques have further accelerated our understanding of the microbial community structure within wastewater treatment systems. Presently, many crucial processes in wastewater treatment systems rely on uncultured prokaryotic organisms. This development has opened up vast opportunities for the exploration of essential microorganisms. Fundamental research has indicated that the diversity of functionally significant prokaryotic groups in wastewater treatment systems can be influenced by factors such as plant design and shifts in process stability. The presence of contaminants like heavy metals, hydrocarbons, phosphorus, and nitrogen in discharged wastewater and distributed water remains an enduring area of concern and investigation.

ACKNOWLEDGMENT

The authors are grateful to the UGC-DAE-CSR, through a Collaborative Research Scheme (CRS) project, Indore, Madhya Pradesh for the financial support (CRS/2021–22/01/461). This work was partially carried out using the facilities of UGC-DAE CSR, Indore, and DST–FIST Centralized laboratory, Muthayammal College of Arts and Science, Rasipuram, Namakkal Dt. Tamil Nadu, India.

REFERENCES

Abriouel, H., Franz, C.M., Omar, N.B., & Gálvez, A. (2011). Diversity and applications of Bacillus bacteriocins. *FEMS Microbiology Reviews*, 35(1), 201–232.

Absar, A.K., Lehr, J.H., & Keeley, J. (eds) Water and Wastewater Properties and Characteristics. In *Water Encyclopedia: Domestic, Municipal and Industrial Water Supply and Waste Disposal*. John Wiley and Sons, Inc., Hoboken, NJ. 903–905 (2005).

Agnieszka, C.K., & Magdalena, Z. (2016). Bacterial communities in full-scale wastewater treatment systems. *World Journal of Microbiology and Biotechnology*, 32, 66.

Akpor, O.B., Adelani-Akande, T.A., & Aderiye, B.I. (2013). The effect of temperature on nutrient removal from wastewater by selected fungal species. *International Journal of Current Microbiology and Applied Sciences*, 2(9), 328–340.

Akpor, O.B., Ogundeji, M.D., Olaolu, D.T., & Aderiye, B.I. (2014). Microbial roles and dynamics in wastewater treatment systems: An overview. *International Journal of Pure & Applied Bioscience*, 2(1), 156–168.

Al Atrouni, A., Joly-Guillou, M.L., Hamze, M., & Kempf, M. (2016). Reservoirs of non-baumannii Acinetobacter species. *Frontiers in Microbiology*, 7, 49.

Amaral, A.L., Da Motta, M., Pons, M.N., Vivier, H., Roche, N., Mota, M., & Ferreira, E.C. (2004). Survey of Protozoa and Metazoa populations in wastewater treatment plants by image analysis and discriminant analysis. *Environmetrics: The Official Journal of the International Environmetrics Society*, 15(4), 381–390.

APHA (2012). *Standard Methods for the Examination of Water and Wastewater*, 21st edition. APHA, Washington DC.

Baduru, L.K., & Sai, G.D.V. (2015). Effective role of indigenous microorganisms for sustainable environment. *Biotechnology, 5*, 867–876.

Benoit, P.C., Weixiao, Q., Huijuan, L., Beat, M., & Michael, B. (2012). Sources and pathways of nutrients in the semi-arid region of Beijing - Tianjin, China. *Environmental Science & Technology, 46*(10), 5294–5301.

Caccio, S.M., De Giacomo, M., Aulicino, F.A., & Pozio, E. (2003). Giardia cysts in wastewater treatment plants in Italy. *Applied and Environmental Microbiology, 69*(6), 3393–3398.

Chun, H.J., Fang, W., Zhen, Y.Y., Ping, X., Hong, J.K., Hong, W.L., Yi, Y.Y., & Jian, H.G. (2015). Study on screening and antagonistic mechanisms of *Bacillus amyloliquefaciens* 54 against bacterial fruit blotch (BFB) caused by *Acidovorax avenae* subsp. *Citrulli*. *Microbiological Research, 170*, 95–104.

Cornel, P., & Krause, S. (2008). Membrane Bioreactors for Wastewater Treatment (Chapter 9). In *Advanced Membrane Technology and Applications*. John Wiley and Sons, Inc., Hoboken, NJ. https://doi.org/10.1002/9780470276280.ch9.

Decicco, B.T. (1979). *Removal of Eutrophic Nutrients from Wastewater and Their Bioconversion to Bacterial Single Cell Protein for Animal Feed Supplements, Phase II*. University of District of Columbia, Water Resources Research Center.

Deepak, B., Ansari, M.W., Ranjan, K.S., & Narendra, T.B. (2014). Biofertilizers function as key player in sustainable agriculture by improving soil fertility, plant tolerance and crop productivity. *Microbial Cell Factories, 13*, 66.

Della Rocca, C., Belgiorno, V., & Meriç, S. (2007). Overview of in-situ applicable nitrate removal processes. *Desalination, 204*(1–3), 46–62.

Department of Natural Science (2006). Wastewater characterization for evaluation of biological phosphorus removal. www.dnr.state.wi.us/org/water. Accessed 13 Jun 2006.

Department of Water Affairs and Forestry (1993). *South African Water Quality Guidelines: Volume 3-Industrial use*. Department of Water Affairs and Forestry.

Díaz, S., Martin-Gonzalez, A., & Gutierrez, J.C. (2006). Evaluation of heavy metal acute toxicity and bioaccumulation in soil ciliated protozoa. *Environment International, 32*(6), 711–717.

Diep, C.N., & Cuc, N.T.K. (2013). Heterotrophic nitrogen removal bacteria in sedimentary and water of striped catfish ponds in the Mekong Delta, Vietnam. *American Journal of Life Sciences, 1*(1), 6–13.

Djadouni, F. (2019). Microorganisms based biological agents in wastewater treatment and agriculture: Potential use and benefits. http://hdl.handle.net/11452/20650.

Djadouni, F., & Madani, Z. (2016). Recent advances and beneficial roles of Bacillus megaterium in agricultures and others fields. *International Journal of Biology, Pharmacy and Applied Sciences, 5*(12), 3160–3173.

Eikelboom, D.H. (2000). *Process Control of Activated Sludge Plants by Microscopic Investigation*. IWA Publishing, London.

EPA (1996). U.S. Environmental Protection Agency, American Society of Civil Engineers, and American Water Works Association technology transfer handbook: management of water treatment plan residuals EPA/625/R-95/008. Office of Research and Development, Washington, DC, p. 281.

Ferrera, I., & Sanchez, O. (2016). Insights into microbial diversity in wastewater treatment systems: How far have we come?. *Biotechnology Advances, 34*(5), 790–802.

Fried, J., & Lemmer, H. (2003). On the dynamics and function of ciliates in sequencing batch biofilm reactors. *Water Science and Technology, 47*(5), 189–196.

Gerardi, M.H. (2007). The protozoa puzzle. Water and Wastewater Products.

Gopinath, S.M., Ismail, S.M., & Ashalatha, A.G. (2015) Bioremediation of lubricant oil pollution in water by Bacillus megaterium. *International Journal of Innovative Research in Science, Engineering and Technology, 4*(8), 6773–6780.

Gray, F.N. (2002). *Water Technology: An Introduction for Environmental Scientists and Engineers.* Butterworth-Heinemann, Oxford, pp. 35–80.

Gray, N.F. (2000). An introduction for environmental scientists and engineers. *Water Technology* (3rd ed., pp. 194–231).

Hamdy, A., Shatanawi, M., & Smadi, H. (2005). Urban wastewater problems, risks and its potential use for irrigation. *The use of non-conventional water resources. Bari: CIHEAM/EU-DG Res,* 15–44.

Hussein, H., Farag, S., Kandil, K., & Moawad, H. (2005). Tolerance and uptake of heavy metals by Pseudomonads. *Process Biochemistry, 40*(2), 955–961.

Haq, I., & Kalamdhad, A.S. (2021). Phytotoxicity and cyto-genotoxicity evaluation of organic and inorganic pollutants containing petroleum refinery wastewater using plant bioassay. *Environmental Technology & Innovation, 23,* 101651.

Haq, I., & Kalmdhad, A.S. (2023). Enhanced biodegradation of toxic pollutants from paper industry wastewater using Pseudomonas sp. immobilized in composite biocarriers and its toxicity evaluation. *Bioresource Technology Reports, 24,* 101674.

Haq, I., Kalamdhad, A.S., & Pandey, A. (2022). Genotoxicity evaluation of paper industry wastewater prior and post-treatment with laccase producing *Pseudomonas putida* MTCC 7525. *Journal of Cleaner Production, 342,* 130981.

Haq, I., Kumar, S., Raj, A., Lohani, M., & Satyanarayana, G.N.V. (2017). Genotoxicity assessment of pulp and paper mill effluent before and after bacterial degradation using *Allium cepa* test. *Chemosphere, 169,* 642–650.

Haq, I., Kumar, S., Kumari, V., Singh, S.K., & Raj, A. (2016a). Evaluation of bioremediation potentiality of ligninolytic *Serratia liquefaciens* for detoxification of pulp and paper mill effluent. *Journal of Hazardous Materials, 305,* 190–199.

Haq, I., Kumari, V., Kumar, S., Raj, A., Lohani, M., & Bhargava, R.N. (2016b). Evaluation of the phytotoxic and genotoxic potential of pulp and paper mill effluent using *Vigna radiata* and *Allium cepa. Advanced Biology, 2016,* 1–10, Article ID 8065736.

Haq, I., & Raj, A. (2018). Markandeya Biodegradation of Azure-B dye by *Serratia liquefaciens* and its validation by phytotoxicity, genotoxicity and cytotoxicity studies. *Chemosphere, 196,* 58–68.

Jayakumar, P., & Natarajan, S. (2012). Microbial diversity of vermin compost bacteria that exhibit useful agricultural traits and waste management potential. *Springer Plus,* 1, 26.

Jenkins, D., Richard, M.G., & Daigger, G.T. (2004). *Manual on the Causes and Control of Activated Sludge Bulking and Foaming and Other Solids Separation Problems.* Lewis Publishers, Washington, DC.

Jimenez-Cisneros, B. E. (2007). Helminth ova control in wastewater and sludge for agricultural reuse. *Water and Health,* 2, 1–12.

Karthiga Devi, K., & Natarajan, K.A. (2015). Isolation and characterization of a bioflocculant from Bacillus megaterium for turbidity and arsenic removal. *Mining, Metallurgy & Exploration, 32*(4), 222–229.

Khaled, S., & Balkhair, S. (2016). Microbial contamination of vegetable crop and soil profile in arid regions under controlled application of domestic wastewater. *Journal of Biological Sciences, 23*(1), S83–S92.

Khan, J.A. (2011). Biodegradation of Azo dye by moderately halotolerant *Bacillus megaterium* and study of enzyme azoreductase involved in degradation. *Advanced Biotechnology, 10,* 21–27.

Kone, D., Cofie, O., Zurbru, C., Gallizzi, K., Moser, D., Drescher, S., & Strauss, M. (2007). Helminth eggs inactivation efficiency by fecal sludge dewatering and co-composting in tropical climates. *Water Research,* 43, 97–4402.

Kris, M. (2007). Wastewater pollution in China. Available from http://www.dbc.uci/wsustain/suscoasts/krismin.html. Accessed, 16(06), 2008.

Kumar, A., Prakash, A., & Johri, B.N. (2011). Bacillus as PGPR in crop ecosystem. In *Bacteria in Agrobiology: Crop Ecosystems* (pp. 37–59). Springer, Berlin, Heidelberg. https://doi.org/10.1007/978-3-642-18357-7_2.

Larsdotter, K. (2006). *Microalgae for phosphorus removal from wastewater in a Nordic climate* (Doctoral dissertation, KTH).

LeChevallier, M.W., & Au, K. (2004). Inactivation (disinfection) processes. In: *Water Treatment and Pathogen Control*. Mark W LeChevallier and Kwok-Keung Au (editors). IWA Publishing, London, 41–65.

Liang, L., Zhigang, Z., Xiaoli, H., Xue, D., Changan, W., Jinnan, L., Liansheng, W., & Qiyou, X. (2016). Isolation, identification and optimization of culture conditions of a bioflocculant-producing bacterium Bacillus megaterium SP1 and its application in aquaculture wastewater treatment. *BioMed Research International*, 25, 9.

Lloyd, B.J., & Frederick, G.L. (2000). Parasite removal by waste stabilisation pond systems and the relationship between concentrations in sewage and prevalence in the community. *Water Science and Technology*, 42(10–11), 375–386.

Lofrano, G., Meriç, S., Zengin, G.E., & Orhon, D. (2013). Chemical and biological treatment technologies for leather tannery chemicals and wastewaters: A review. *Science of the Total Environment*, 461, 265–281.

Mara, D.D., Millw, S.W., Pearson, H.W., & Alabaster, G.P. (2007). Waste stabilization ponds: A viable alternative for small community treatment systems. *Water and Environment Journal*, 6, 74.

Mara, D., & Horan, N.J. (Eds.). (2003). *Handbook of Water and Wastewater Microbiology*. Elsevier, Burlington.

Metcalf, X., & Eddy, X. (2003). Wastewater Engineering: Treatment and Reuse. In: *Wastewater Engineering, Treatment, Disposal and Reuse*. Tchobanoglous, G., Burton, F.L., Stensel, H.D. (editors). 4th edition, Tata McGraw-Hill Publishing Company Limited, New Delhi, India.

Mingjun, S., Yanqiu, W., & Xue, S. (2009). Study on bioremediation of eutrophic lake. *Journal of Environmental Sciences*, 21, S16–S18.

Mittal, A. (2011). Biological wastewater treatment. *Water Today*, 1, 32–44.

Molla, A.H., Fakhru'l-Razi, A., & Alam, M.Z. (2004). Evaluation of solid-state bioconversion of domestic wastewater sludge as a promising environmental-friendly disposal technique. *Water Research*, 38(19), 4143–4152.

Momba, M.N.B., & Mfenyana, C. (2005). Inadequate treatment of wastewater: a source of coliform bacteria in receiving surface water bodies in developing countries-case study: Eastern Cape Province of South Africa. *Water Encyclopedia*, 1, 661–667.

Nasr, M., Elreedy, A., Abdel-Kader, A., Elbarki, W., & Moustafa, M. (2014). Environmental consideration of dairy wastewater treatment using hybrid sequencing batch reactor. *Sustainable Environment Research*, 24(6), 449–456.

Nelson, K., Cisneros, B.J., Tchobanoglous, G., & Darby, J. (2004). Sludge accumulation, characteristics, and pathogen inactivation in four primary waste stabilization ponds in central Mexico. *Water Research*, 38(1):111–127.

Ng, K.K., Shi, X., Ong, S.L., & Ng, H.Y. (2016). Pyrosequencing reveals microbial community profile in anaerobic bio-entrapped membrane reactor for pharmaceutical wastewater treatment. *Bioresource Technology*, 200, 1076–1079.

Odonkor, S.T., & Ampofo, J.K. (2013). Escherichia coli as an indicator of bacteriological quality of water: An overview. *Microbiology Research*, 4(1), e2.

Oehmen, A., Lemos, P.C., Carvalho, G., Yuan, Z., Keler, J., Blackall, L.L., & Reis, A.M. (2007). Advances in enhanced biological phosphorus: From micro to macro scale. *Water Research*, 41, 2271–2300.

Oh, S.T., Kim, J.R., Premier, G.C., Lee, T.H., Kim, C., & Sloan, W.T. (2010). Sustainable wastewater treatment: How might microbial fuel cells contribute. *Biotechnology Advances*, 28(6), 871–881.

Okoh, A.I., Odjadjare, E.E., Igbinosa, E.O., & Osode, A.N. (2007). Wastewater treatment plants as a source of microbial pathogens in receiving watersheds. *African Journal of Biotechnology*, 6(25). https://doi.org/10.5897/AJB2007.000-2462.

Paillard, D., Dubois, V., Thiebaut, R., Nathier, F., Hoogland, E., Caumette, P., & Quentin, C. (2005). Occurrence of Listeria spp. in effluents of French urban wastewater treatment plants. *Applied and Environmental Microbiology*, 71(11), 7562–7566.

Pena-Varon, M.R. (2002). *Advanced primary treatment of domestic wastewater in tropical countries: development of high rate anaerobic ponds* (Doctoral dissertation, University of Leeds (School of Civil Engineering)).

Radjenovic, J., Matosic, M., Mijatovic, I., Petrovic, M., & Barcelo, D. (2007). Membrane bioreactor (MBR) as an advanced wastewater treatment technology. In: *Emerging Contaminants from Industrial and Municipal Waste*, 37–101. https://doi.org/10.1007/9 78-3-540-79210-9_2.

Rein, M. (2005). Copigmentation reactions and color stability of berry anthocyanins.

Richard, M., Brown, S., & Collins, F. (2003, June). Activated sludge microbiology problems and their control. In: *20th Annual USEPA National Operator Trainers Conference*, Buffalo, New York (Vol. 8, pp. 1–21).

Richardson, A.E., Barea, J.M., McNeill, A.M., & Prigent-Combaret, C. (2009). Acquisition of phosphorus and nitrogen in the rhizosphere and plant growth promotion by microorganisms. *Plant and Soil*, 321(1), 305–339.

Rosenwinkel, K.H., Austermann-Haun, U., & Meyer, H. (2005). Industrial wastewater sources and treatment strategies. In: *Environmental Biotechnology: Concepts and Applications*. https://doi.org/10.1002/3527604286.ch2.

Sanguinetti, G., Tortul, C., García, M., Ferrer, V., Montangero, V., & Strauss, M. (2005). Investigating Helminth eggs and Salmonella sp. in stabilization ponds treating seepage. *Water Science & Technology*, 51(12), 239–247.

Santamaria, J., & Toranzos, G.A. (2003). Enteric pathogens and soil: A short review. *International Microbiology*, 6(1), 5–9.

Schmidt, I., Sliekers, O., Schmid, M., Cirpus, I., Strous, M., Bock, E., … & Jetten, M.S. (2002). Aerobic and anaerobic ammonia oxidizing bacteria-competitors or natural partners? *FEMS Microbiology Ecology*, 39(3), 175–181.

Schultz, T.E. (2005). Biological wastewater treatment: the use of microorganisms to remove contaminants from wastewater is effective and widespread. To choose the right system from the many options offered, understand the various techniques available and evaluate them based on your requirements. *Chemical Engineering*, 112(10), 44–51.

Shaikh, Z.A., Graham, D.W., & Dolfing, J. (2013). Wastewater treatment: Biological. pp. 2645–2655. 10.1201/9781003045045-61.

Shakir, E., Zahraw, Z., & Al-Obaidy, S. (2017). Environmental and health risks associated with reuse of wastewater for irrigation. *Egyptian Journal of Petroleum*, 26(1), 95–102.

Shilton, A., & Harrison, J.H. (2003). *Guidelines for the hydraulic design of waste stabilisation ponds*. Institute of Technology and Engineering, Massey University.

Shumaila, S., Saira, A., & Abdul, R. (2013). Characterization of cellulose degrading bacterium, *Bacillus megaterium* S3, isolated from indigenous environment. *Pakistan Journal of Zoology*, 45(6), 1655–1662.

Singhania, R.R., Christophe, G., Perchet, G., Troquet, J., & Larroche, C. (2012). Immersed membrane bioreactors: An overview with special emphasis on anaerobic bioprocesses. *Bioresource Technology*, 122, 171–180.

Stevik, T.K., Aa, K., Ausland, G., & Hanssen, J.F. (2004). Retention and removal of pathogenic bacteria in wastewater percolating through porous media: A review. *Water Research*, 38(6), 1355–1367.

Stottmeister, U., Wießner, A., Kuschk, P., Kappelmeyer, U., Kästner, M., Bederski, O., ... & Moormann, H. (2003). Effects of plants and microorganisms in constructed wetlands for wastewater treatment. *Biotechnology Advances*, *22*(1–2), 93–117.

Sutton, P.M. (2006). Membrane bioreactors for industrial wastewater treatment: Applicability and selection of optimal system configuration. *Proceedings of the Water Environment Federation*, *2006*(9), 3233–3248.

Swiontek Brzezinska, M., Jankiewicz, U., Burkowska, A., & Walczak, M. (2014). Chitinolytic microorganisms and their possible application in environmental protection. *Current Microbiology*, *68*(1), 71–81.

Tchobanoglus, G., Burton, F., & Stensel, H.D. (2003). Wastewater engineering: Treatment and reuse. *American Water Works Association, Journal*, *95*(5), 201.

Templeton, M.R., & Butler, D. (2011). *Introduction to Wastewater Treatment*. Bookboon, Copenhagen.

Tingting, Z., & Dittrich, M. (2016). Carbonate precipitation through microbial activities in natural environment, and their potential in biotechnology: A review. *Frontiers in Bioengineering & Biotechnology*, *4*, 4.

Tom, S., Simon, J., Bruce, T., & Keith, B., (2000). *Membrane Bioreactors for Wastewater Treatment*. IWA Publishing, London.

Tortora, G.J., Funke, B.R., & Case, C.L. (2010). *Microbiology An Introduction*, 10th edition. Benjamin Cummings, San Francisco, CA.

United Nations Environmental Programme (2012). Wastewater and stormwater Treatment.

US-EPA, (2000) "Oxidation Ditches" Wastewater Technology Fact Sheet Washington DC.

Van Haandel, A., & Van Der Lubbe, J. (2007). *Handbook Biological Waste Water Treatment-Design and Optimisation of Activated Sludge Systems*. Webshop Wastewater Handbook.

Wang, Y.H., Dong, B., Mao, Y.L., & Yan, Y.S. (2010). Bioflocculant-producing Pseudomonas alcaligenes: Optimal culture and application. *Huanjing Kexue yu Jishu*, *33*(3), 68–71.

Ward-Paige, C.A., Risk, M.J., & Sherwood, O.A. (2005). Reconstruction of nitrogen sources on coral reefs: δ15N and δ13C in gorgonians from Florida Reef Tract. *Marine Ecology Progress Series*, *296*, 155–163.

Water Environment Association, Activated sludge, Manual of Practice No. 9 (1987).

World Health Organization. (2006). *WHO Guidelines for the Safe Use of Wasterwater Excreta and Greywater* (Vol. 1). World Health Organization, Geneva.

Yazdani, M., Bahmanyar, M.A., Pirdashti, H., & Esmaili, M.A. (2009). Effect of phosphate solubilization microorganisms (PSM) and plant growth promoting rhizobacteria (PGPR) on yield and yield components of corn (*Zea mays* L.). *World Academy of Science, Engineering and Technology*, *49*(1), 90–92.

10 Microbes-Assisted Development of Biofertilizer from Crop Stubble

Anfal Arshi, Suchita Atreya, Jyotsana Maura, Anshu Singh, and Izharul Haq

10.1 INTRODUCTION

On a local and regional scale, burning of agriculture waste pollutes land and water severely. According to Purohit and Chaturvedi (2018), burning paddy straw is predicted to cause 3.85 million tons of nutritional losses, including 59,000 tons of nitrogen, 20,000 tons of phosphorus, 35,000 tons of potassium, and organic carbon. This has a negative impact on the soil's nutrient budget (Benbi et al., 2006). It produces smoke, which, when combined with other gases in the atmosphere such as methane, nitrogen oxide, and ammonia, can generate severe pollution.

One such input that has emerged as the most widely used tool in the rice–wheat cropping system is the combined automated harvester. Combination harvesters are being used much more frequently. Over 80% of the rice crop is harvested with the help of this device. However, using a combination harvester has actually had an impact. Agriculture residue management became more of a challenge. Due to the massive amount of rice residue left in the open fields by the use of combined harvesters, these fields must be burned (Hegde, 2010). The combined harvester scatters the rice residue over the fields, making it more difficult to collect. In general, farmers believe that burning rice stubble is the most expedient and affordable method of disposal. Furthermore, there isn't much of a choice for farmers other than to burn it because there isn't enough time to plant the wheat crop once the rice crop is harvested.

Each year, over 15 million tons of paddy straw are generated. Various estimates indicate that each year 7–8 million tons of rice residue are burned in open fields. Burning rice waste has a variety of negative effects both on and off the farm, such as losses in soil nutrients and soil organic matter, production and productivity, air quality, biodiversity, water and energy efficiency, as well as impacts on human and animal health. In India, air pollution from residue burning can be very dangerous to people's health, causing or making a number of health issues worse, and increasing the risk of serious car accidents by lowering visibility.

DOI: 10.1201/9781003368120-10

One of the known obstacles to the long-term sustainability of the rice–wheat agricultural system is the loss of soil organic matter as a result of burning. Field-harvested straw is highly valuable economically as fuel, food for animals, and raw material for manufacturing. Paddy straw is fed to animals in southern India, while wheat straw is preferred in the north (Hegde, 2010). The rice–wheat cropping system's residue can be put to a number of uses, but only if it is taken out of the field and separated from the grain. Burning reduces animal access to straw, which is already scarce, by more than 40%. However, when the majority of the residue is left on the field after combine harvesting for burning, the viability of the rice–wheat cropping system as a whole is put in jeopardy (Thakur, 2003). Many farms are now operating with no tillage after stubble burning. 10% of the total land was seeded with wheat in 2005–2006 using zero-till equipment. According to reports, less than 1% of farmers use paddy straw because burning it would need more tillage (Singh et al., 2008). Alternatives for managing crop residues could include creating methods for extracting residue from pastures and using it as industrial or animal feed. Increased machine-harvested straw decomposition can lead to better soil nutrients.

A novel method for increasing their digestibility is biological treatment of agricultural leftovers. There are relatively few groups of microorganisms that can manufacture the ligninolytic enzymes, despite the fact that cellulolytic fungi can be found in all major fungal taxa. As saprophytism is the most common way of life for soil microorganisms, saprophytic fungi are essential to the sustainability of the arable soil ecosystem in terms of nutrient turnover. Fungi are the primary contributors to the formation of dissolved organic matter due to their role in the partial decomposition of organic matter. The use of microbial sprays that can speed decomposition of residue is also an option. Further research on inorganic nitrogen and its negative effects owing to nitrogen deficit is required before choosing to plant in residual.

The humus content of the soil and the biological constitute on of plant leftovers, as well as the rate at which they decompose. With the addition of plant residues, the aggregate stability started by microbial decomposition of the carbohydrate and amino acid content of the residue is then strengthened by interaction with phenolic acids like vanillin or vanillic acid. This done by analyzing the rate of decomposition of corn (*Zea mays* L.), soybean (*Glycine max* L.), alfalfa (*Medicago sativa* L.), oat (*Avena sativa*)

Recycling essential plant elements like nitrogen, phosphorous, and sulfur—all of which are present in organic waste—is particularly important. According to McMurtrie et al. (2012), because nutrient intake is correlated with photosynthetic activity, particularly during plant development, the biomass accumulation model can predict a plant's nutritional demand. It's crucial to time the release of nutrients from organic wastes with the requirement of a plant during the growing season. Mineral fertilizers may not be as necessary if organic leftovers provide enough nutrients for a plant's growth.

10.2 PRODUCTION OF CROP STUBBLE

The demand for and production of food have increased along with the global population. A huge amount of agricultural residues are now there as a result. Agricultural residues are often divided into two categories: crop residues and agro-industrial residues. Plant remains that are still visible in the field after a crop has been harvested are known as crop residues. Crop wastes include things like wheat shafts, shoots, and straw. In contrast, agro-industrial residues include leftovers from the food processing industry as well as postharvest processes like washing, screening, and milling. The agricultural waste generated each year is made up of straw, roots, shafts, and other portions of corn, wheat, and rice, which account for 40.6%, 24.2%, and 15.7%, respectively (Hegde, 2010). For instance, it was projected that China will produce over 807 million tons of agricultural residue overall in 2009, up from an average yield of 716 million tons for the ten years prior. According to the Ministry of New and Renewable Energy, India produces over 500 million tons (Mt) of agricultural waste per year, largely from grains. The highest producing states are Punjab (51 Mt), Maharashtra (60 Mt), and Uttar Pradesh (60 Mt) (46 Mt).

The substantial production of lignocellulose goods like straw and stalks reduces potential environmental risks. For instance, tillage and harvesting all agricultural waste can contribute to soil erosion and a reduction in the soil's organic matter content. The binding of protein molecules to the phenolic components produced in anaerobic soils can result in the fixation of organic matter to nitrogen (N) (Thakur, 2003). Furthermore, some research has indicated that employing fresh materials as a soil improver may have negative effects on the stability and maturation of organic matter. Utilizing organic waste, such as agricultural wastes, is one of these problems. As a result of protracted microbial oxidation, one of these outcomes is the development of anaerobic conditions and enhanced rates of organic carbon mineralization in native soils. Due to the creation of ammonia, ethylene oxide, and organic acids, these new residues may also be phytotoxic. However, when they burn, the air quality is lowered. Consequently, an environmentally friendly treatment approach is required (Thakur, 2003).

10.3   CHEMICAL COMPOSITION OF RICE AND WHEAT STUBBLE

Crop waste is a valuable natural resource, not a waste. Crop residues are excellent nutritional sources because they retain about 25% of the nitrogen (N), phosphorus (P), 50% of the sulfur (S), and 75% of the potassium (K) that cereal crops absorb. Experts calculated the amount of nutrients present in rice (Hegde, 2010). Paddy straw contains 11 kg/ha S, 140 kg/ha K, 6 kg/ha P, and 39 kg/ha N. It becomes Rs. 424/ha of N, Rs. 96/ha of P, and Rs. 231/ha of S when it is converted into monetary values, for a total of Rs. 751/ha.

10.4   AGRICULTURAL RESIDUES TRANSFORMATION
AND STABILIZATION

Crop waste management, as well as its right use and disposal, have grown to be serious issues for human society. Due to the rise of agricultural waste, landfill space has become a prohibitive barrier for disposal. Processing residues and recycling them so become practical options for their disposal. Since soil organic matter (OM) is known to improve damaged soils and provide crop nutrients, agricultural leftovers (such as crop residues and animal manures) have been applied to soil for decades as an intriguing environmental and agricultural method for maintaining and enhancing OM. It is therefore generally accepted that stabilizing these residues before application to soil could help increase their residence period by reducing CO_2 oxidation in soil agro-ecosystems. There are several biological breakdown stages that the agricultural residue goes through. The crop residue's carbon is broken down by soil microbes, but they also require nitrogen to do so (Bridgham and Ye, 2015). However, if the C concentration is higher than the N concentration, it will take longer for the soil bacteria to decompose the organic material and they will use the soil N to finish the job. Several studies have shown that using microbial consortia for straw treatment is another effective and strategic method.

Managing agricultural waste through composting is a well-known and effective technology. Better soil health and fertility, which increase agricultural output and soil biodiversity, are just a few benefits that these leftovers offer. The majority of (macro)fungi are saprophytic, which means they recycle and decompose dead plant and animal debris. Fungi are common, morphologically similar creatures found in a variety of habitats. The connection between biotic variety and ecosystem function has drawn considerable attention from ecologists. Westerdijk created the word "association" to describe the distinctive combination of microfungi on decomposing organic substrates. The ability of each species to decompose vary depending on environmental factors and interactions with other fungi. Although many fungal species can degrade the majority of substrates, this is not true of all of them. It is understood that the existence of some taxa is dependent on the kind and quality of litter available, but there hasn't been much research on the relationship between different saprophytic species and particular types of soil.

For instance, cereal stubble contains *Stemphylium* spp., *Alternaria tenuis*, and *Cladosporium herbarum*. *Chaetomium, Myrothecium, Trichoderma, Fusarium, Aspergillus, Penicillium*, and *Trichonympha* are a few of the fungi that contribute to decomposition (Henriksen and Breland, 2002). In another study, it was asserted that a microbial consortium made up of *Aspergillus nidulans, Trichoderma viride, A. awamori*, and *Phanerochaete chrysosporium* was produced during the composting of different crop residues with chicken manure and rock phosphate to produce nitrogen-enriched phosphocompost in a short amount of time. *A. nidulans, Scytalidium thermophilum*, and *Humicola* sp. were shown to be quite helpful and advantageous in the breakdown of soybean waste and rice straw. It is believed that the ligno-cellulolytic microbe is a technologically associated strategy that can increase the capacity for degradation of agricultural wastes. It can also be recommended for the decomposition of straw and the creation of high-quality compost that contains more macronutrients. The most fundamental type of group among microbiological agents is composed of fungi, which exhibit rapid solid-surface colonization.

It is plausible to suppose that a fungus' or a fungal community's capacity to produce enzymes that can break down a substrate's cell wall components is a prerequisite for that organism's or community's capacity to breakdown a substrate. One of the key factors in the effective colonization of a dead organic substrate by the fungi is "excellent enzyme-producing equipment," which is crucial for the invasion of plant tissue by fungi. The "pace at which it can permeate successive cell walls in a cellular tissue" is another crucial quality. Examining the relationships between the species occupying the substrate will be useful in determining the bioinoculant to be used for the environmental breakdown of organic waste. The lignin, cellulose, and chitin in this substance are broken down and digested by saprophytic fungi into simple soluble compounds that they and plants may take as food. They aid in the recycling of vital nutrients, especially carbon and nitrogen, and the elimination of the buildup of dead OM.

Cellulose, lignin, and hemicelluloses make up the majority of crop tissues. All vascular plants' cell walls and tissues are strengthened and stiffened by the complex aromatic macromolecule lignin, which serves as a glue between polysaccharide filaments and fibers. Chemically, lignin is a heterogeneous molecule that is linked by many covalent bonds. Unlike most other natural polymers (cellulose, starch, proteins, etc.), which hydrolytic enzymes can usually break down, these linkages are insoluble to them. Lignin is less prevalent in the composting process than cellulose or hemicelluloses. Future efforts to hasten the conversion and humification of lignin-rich wastes may involve the addition of organisms like saprophytic fungus that can operate on lignin's chemical linkages. The functions of saprophytic fungi as primary, secondary, and tertiary decomposers in natural settings are well established. Additionally, fungi are the most effective breakers of lignin, hemicellulose, and celluloses, as well as humic substances (HS). Due to the fact that bacteria cannot digest many of these chemicals, fungi are assumed to be more effective than bacteria. Recent studies have discovered that some isolates are better at degrading lignin than others, and extracellular enzymes like peroxidases and laccase seem to be involved. Extracellular enzymes such as manganese peroxidase (MnP), laccase (LAC), and lignin peroxidase (LiP) are secreted by saprophytic fungus (Henriksen and Breland, 2002).

Specific basidiomycetous fungus are the only ones that can produce the peroxidases manganese (MnP) and lignin (LiP). As ecophysiological groups capable of secreting these enzymes, white-rot and litter decomposers (WRF and LDF) have been identified. When saprophytic fungi create these enzymes, lignin breakdown (creating the so-called ligninolytic system) leads in the formation of unstable molecules, such as phenoxy radicals, which can subsequently condense and polymerize. Each enzymatic product is likely to be influenced by the reaction circumstances, including pH, humidity, percentage oxygen, and electrical conductivity, in addition to the enzymes and substrates used. Cellulases, hemicellulases, and esterases, as well as other enzymes that depolymerize cellulose, hemicellulose, and glucosides, are believed to be important in the breakdown of lignocellulosic biomass. Half of the approximately 8,500 species of lignocellulose-degrading saprotrophs, or basidiomycetes, have been found in soils, wood, and plant litter. These fungi can also break down refractory substances in addition to HS. Due to their enormous size and stability, HS are unlikely to be digested by microbial cells intracellularly; instead, they seem to be eliminated by non-specific extracellular enzymes.

WRF and LDF only convert and break down macromolecules in the presence of additional readily biodegradable C sources. Because of their extracellular enzymatic activity, white-rot fungi (WRF) and litter decomposer fungi (LDF) are frequently used to break down the lignocellulosic components of agricultural residues (crop residues) and speed up the process of stabilization into humic compounds (Henriksen and Breland, 2002). Examples include WRF, which is widely known for its capacity to biodelignify wood particles, wheat straw and bamboo sticks.

Different fungal taxa predominate at various stages of degradation in various environments. It is unknown, however, what biological traits of fungi are responsible for this pattern. A number of mycologists looked into the well-established hypothesis that different fungal phyla prefer distinct stages of decay, and that fungal succession is impacted by a species' evolutionary history. The goal would have been to identify the bioinoculant that should be used to accelerate the breakdown of different types of garbage in order to protect the environment and ensure public safety. This could be useful in choosing the right inoculants to slow down the rate at which postharvest crop residues decompose for long-term resource usage in environmentally friendly agro farming techniques.

In order to extract the fungi required to break down crop stubble, dilution plate technique is performed on a laboratory scale. Particular antibiotics must be introduced to the medium in order to stop bacterial growth. It is then possible to inoculate the isolated fungi into flasks filled with agricultural stubble. The minimum number of replicas for each organism is three. The flask must be forcefully shaken after inoculation and incubated at the ideal temperature. Degraded stubbles should be taken out of flasks in triplicate at least once every month for at least 3 months, and various plant tissue components such lignin, cellulose, nitrogen, soluble nitrogen, and soluble carbohydrates should be measured. In order to promote soil, using mycoflora as a biofertilizer instead of burning may show to be more successful. Additionally, numerous studies have demonstrated that employing mixed cultures of microbes is preferable to using single organisms. This is clearly demonstrated by the fact that some of the organisms in the mixed cultures help with the first conversion of agricultural residue while others carry out secondary conversion, enhancing the effectiveness of biodegradation.

10.5   MICROBIOLOGICAL MINERALIZATION OF CROP RESIDUES

Mineralization is the biological breakdown of complex organic molecules into more basic inorganic ones. They are reduction and oxidation reactions in which organisms, like many other living organisms, use the energy contained in reduced organic compounds to grow and build their own cells by converting them into more oxidized forms. Through a stoichiometric process, elements like carbon, nitrogen, and phosphorus are supplied to organisms in proportion to the demand placed on them by the constituent parts of their cells (Bridgham and Ye, 2015). Rusnak claims that 20%–25% of agricultural waste is humified, producing soil humus, and indeed the remaining 75%–80% is mineralized into nutrients that are available to plants (Rusnak, 2017).

Crop waste is a good source of OM and can be added back to the ground to recycle nutrients and enhance the physical, chemical, and biological characteristics of the soil. According to Henriksen and Breland, managing agroecosystems requires understanding the dynamics of agricultural residue degradation (2002).

Decomposition of crop waste is considered as a crucial part of the nutrient cycle; it also helps to protect organic compounds and immobilize nutrients in the soil structure. To lessen the risk of erosion and to encourage water retention, crop wastes are sprayed to the soil (Chen et al., 2014). Crop waste is a good source of OM that can be reinserted into the soil to improve its physical, chemical, and biological properties and recycle nutrients. Henriksen and Breland assert that comprehension of the dynamics of agricultural residue degradation is necessary for managing agroecosystems (2002). Crop waste decomposition is regarded as an essential component of the nutrient cycle; it also aids in the preservation of organic components and the immobilization of nutrients in the soil structure. Crop wastes are sprayed into the soil to minimize the risk of erosion and to promote water retention (Chen et al., 2014).

It is challenging to predict how many nutrients will be available to crops at any point of time, because of the intricate processes governing the breakdown of agricultural wastes and the release of nutrition. Additionally, the type of crop residues and how they are handled can have a significant impact on the quantity and quality of OM in the soil, as well as the amount of nutrients that are available to plants (Yadvinder-Singh and Timsina, 2005).

Microorganisms control a variety of crucial ecological processes in the soil ecosystem. They have a crucial role in humic compound formation, nutrient transport, and breakdown in the soil (Jamir et al., 2019; Rashid et al., 2016; Bardgett and van der Putten, 2014). The proportion of proteobacteria, acidobacteria, actinobacteria, and fungi Ascomycota varies with the rate of breakdown of plant waste. During the process, there has been a noticeable decline in the number of bacteria, but an increase in the quantity of fungi (Strickland et al., 2009).

This enhances the surface of mineralized plant waste, making it easier for other microorganisms to colonize it and making hydrolyzed polymer substances more accessible (Abbasi et al., 2015) (cellulose, lignin, protein). Plant leftovers are broken down by successive groups of microorganisms. Chemicals that break down quickly are quickly consumed by bacteria, creating an increase in the refractory component and the fungal community. Proteolytic and cellulolytic enzyme-producing bacteria predominate at first, but as the process goes on, the ratio starts to favor nitrifiers. Fungi that are expert polymer degraders take up simple, soluble components (such as sugars) first in a fungal colony. The most resistant component is broken down by fungi in the last stage of disintegration.

According to a study by Henriksen and Breland (2002), straw has a lower ratio of easily absorbable carbs and proteins than fresh herbaceous material such leftover clover, but it is higher in cellulose and hemicelluloses. Therefore, for breakdown, it is essential for microorganisms that produce extracellular cellulases and hemicellulases to colonize and grow properly. Particularly crucial to this process are moulds. *Eubacteria* from the *Cellulomonas* sp., *Cellvibrio* sp., and *Cytophaga* sp. genera, as well as fungi from the *Trichoderma*, *Fusarium*, and *Mycogone* genera, show strong cellulolytic properties (Galus-Barchan and Chmiel, 2019). *Thermobifida* sp., *Curtobacterium* sp., *Streptomyces* sp., *Clostridium* sp. (especially *C. thermocellum*), and *Flavobacterium* sp. all have cellulolytic characteristics.

One of the most intricate compounds in plant debris that enters the soil is lignin. This aromatic molecule is quite hard to decompose because of its complex structure, high molecular weight, and insolubility. It can only be broken down by bacteria that have the necessary enzymes, such laccase and tyrosinase. In addition to the fungi *Aspergillus* sp., *Penicillium* sp., *Trichoderma* sp., and *Chaetomium* sp., soil bacteria from the families *Azotobacter, Xanthomonas, Pseudomonas,* and *Agrobacterium* were also identified. Basidiomycetes and ascomycetes break down lignin, including species of *Xylaria, Hypoxylon,* and *Libertella* (Galus-Barchan and Chmiel, 2019).

10.5.1 ENVIRONMENTAL ELEMENTS THAT AFFECT THE RATE OF CROP RESIDUE MINERALIZATION

Microbial activity during the breakdown of OM is affected by a wide range of physical, chemical, and biological factors, including soil type, oxygenation, moisture, pH, salinity, the concentrations of OM and trace elements, the presence of stimulating or inhibiting substances, and the presence of other microorganisms. The pace of OM decomposition is influenced by a variety of factors, including the climate, the seasons of the year, agrotechnical practices, fertilizer types, and plant cover. Type, species, even the range and developmental stage of the plant have all been discussed.

Tropical climates generally have a greater rate of mineralization of plant-based OM than temperate climates. This process doubles in speed when the mean annual air temperature rises by 8°C–9°C. Due to the faster rate of plant detritus breakdown in tropical climes, it is difficult to keep an equilibrium in the OM content of soil agro-ecosystems. As a result, substantial amounts of OM must be added to the soil in this climate zone in order to maintain an acceptable level of OM. In milder climates, soils often have more OM because they mineralize more slowly (Pangnakorn et al., 2003).

The rate of mineralization of plant residues increased with soil temperature (from 5°C to 45°C) while looking at the dispersion of rice straw and green manure in soil at various temperatures. Researchers have discovered that temperature is a crucial regulator of the microbial breakdown of crop waste. The study discovered that soil moisture significantly affects the amount of CO_2 produced and released in the soil during the breakdown of plant wastes. The distribution of soybean residues in soil revealed that, as soil temperature and humidity increased, the proportion of nitrogen released from plant residues increased linearly.

Another factor that influences the decomposition of crop waste is soil moisture. The rate of organic nitrogen mineralization and immobilization is significantly influenced by the presence or absence of water on the activity of soil microbes. Numerous interrelated mechanisms limit the activity of soil microorganisms based on the amount of water in the soil. Dry soil has a number of effects, including reduced soluble chemical diffusion, less microbial mobility, and thus limited microbial access to the substrate. Aerobic bacteria's activity may be lowered in damp soil because oxygen transfer is hindered. Due to their less access to organic materials and nutrients that dissolve in water trapped in soil pores, microorganism activity in dried soils is lower (Paul, 2014). On the other hand, in flooded soils, water fills soil pores that are typically filled with oxygen, limiting the microbial population of oxygen. Extremely complicated factors, including soil type and porosity, OM concentration, pH, and soil depth, influence the link between soil water content and microbial activity (Paul et al., 2003).

10.5.2 CROP RESIDUES ARE BENEFICIAL FOR AGRICULTURAL PRODUCTION

Plant detritus affects the texture, number, and structure of soil microbes. It feeds soil microorganisms with nutrients, shields the soil from wind and water erosion, and enhances the hydrological properties and water infiltration of the soil. It raises soil porosity, water retention, heat flow, gas flow, and the availability of macro- and microelements (Brankatschk and Finkbeiner, 2017).

Crop residues added to the soil enhanced fertility, increased cultivation effectiveness, and prevented nutrient loss from the substrate that would have otherwise occurred (Pandiaraj et al., 2015). The treatment of crop residues increased the amount of mineral nitrogen in the soil by 1.22 times, the amount of nitrogen in wheat grain by 1.32 times, and the amount of nitrogen in wheat straw by 1.67 times as compared to the control variation, where no crop residues were treated. According to the study, adding crop wastes to the soil enhanced wheat grain yield by 1.22 times in the first year of the trial and by 1.38 times in the second (Surekha et al., 2003).

The occurrence of soil-borne plant diseases in agricultural operations is reduced as a result of crop residue management. Crop residues in soil may have a direct impact on the viability and activity of plant diseases in the substrate. Arnault et al. (2013) combined laboratory and field studies to evaluate the possible use of compounds derived from leftover onions (*Allium cepa* L.). The study found that onion wastes significantly affected soil-borne diseases due to the higher concentration of disulfides used through the soil. Garlic residues (*A. sativum* L.) also demonstrate many bioactive properties, including inhibitory effects on a variety of soil microorganisms, as a result of the creation of several sulfur-containing chemical compounds that can act upon fungus and bacteria. In conclusion, the complex biochemistry of the sulfur compounds found in allium residues (cysteine, cystine, methionine, and glutathione) that break down and have positive effects on the management of *F. oxysporum* and *Verticillium dahlia* is strongly related to the characteristics of allium residues (Wang et al., 2009). They contain sulfur compounds (cysteine, cystine, methionine, and glutathione), which break down and aid in controlling *V. dahlia* and *F. oxysporum* (Wang et al., 2009).

Some agricultural wastes also impede weed growth (allelopathic activity) by reducing weed access to light, altering soil temperature, and producing chemical compounds (Jones et al., 1999). The results of Chauhan's investigation on how residues from rye (*Secale cereale*) and wheat (*Triticum aestivum* L.) affect the growth of weeds provided support for this hypothesis. To lessen the biomass and density of various weed species, rye is frequently grown as a ground cover crop in soybean and maize plantations. In a maize plantation, rye residues reduced the biomass and density of weeds such common purslane (*Portulaca oleracea* L.), hairy crabgrass (*Digitaria sanguinalis* L.), and lamb's quarters (*Chenopodium album* L.) by more than 75% when compared to a plantation that did not use crop residues. However, some crop wastes, such as wheat, may promote the germination of weed seeds like common wild oats.

10.6 AGRICULTURAL WASTE MANAGEMENT TECHNIQUES

10.6.1 FOR SOIL FERTILITY

Restoring straw to the field, which contains a significant amount of organic carbon, enhances nutrient content in the soil, improving soil fertility and physical and chemical qualities, which is favorable to crop growth. Encourage crop yields to be high. Returning straw to the field has been discovered to have significant benefits for the agricultural ecosystem, according to some researchers. However, research in various climates and soil types have yielded conflicting results, straw incorporation to boost crop output is still a topic of contention. The organic acids released during straw decay are damaging to crops' root systems, and the slower decay of straw will impact subsequent crop roots (Sun et al., 2020).

Soil methane and nitrous oxide emissions account for 42%–59% of anthropogenic emissions globally (Ciais et al., 2013). The straw return was thought to have provided more usable C and N for microorganisms, resulting in an increase in soil N_2O and CH_4 emissions. While the soil is aerobic, methane-oxidizing bacteria and other aerobic microorganisms thrive, while organic soil carbon mineralizes and is primarily removed in the form of CO_2. When the soil is in a deteriorating anaerobic condition, the redox potential of the soil drops, methanogens and other anaerobic microbes in the soil proliferate, and organic soil carbon mineralizes and is mostly eliminated in the form of CH_4. Meanwhile, changes in soil qualities may have an impact on the amount of warming gases released when the straw is returned. When evaluating the carbon and nitrogen storage findings, the overall global warming potential (GWP) of the straw return is larger than that of the typical fertilization technique (Awais et al., 2021; Jin et al., 2020).

The straw return method commonly used by farmers in the production process has many negative effects on the conservation and improvement of soil fertility, the realization of sustainably high crop yields, and eco-environmentally sustainable development due to the lack of a comprehensive straw return theory.

10.7 FEED PRODUCTION

Reusing the nutrients in straw is precisely what it entails. The main therapeutic modalities used today include physical, chemical, and biological ones. Physical approach: shorten the straw and make it softer to increase cattle absorption. Chemical techniques: straw microsilage, straw silage, and straw ammoniation. The withered straw is subjected to a biological process that uses microorganisms or enzymes to break down high-molecular-weight substances like cellulose, hemicellulose, and lignin into lower-molecular-weight ones like monosaccharide oligosaccharide, which are then further transformed to create things like bacterial protein and vitamins. In China, the use of straw as a feed ingredient is quite low, and the majority of feed treatments are still at the conventional silage level. The nutrients in straw are reduced after a succession of treatments, and the amount of ammoniated material is considerable, which is damaging to the environment. These are just two of the numerous unresolved issues with this procedure. As a result, widespread adoption is challenging (Paolini et al., 2018). Before being utilized as animal and poultry feed, livestock dung must first undergo a series of sterilizations due to the presence of various toxic chemicals and nutrients. Animal excrement has a very high concentration of undigested crude protein; in air-dried chicken manure, it is between 24% and 30%; in pig manure, it is between 3.5% and 4.1%; and in cow manure is 1.7%–2.3% (Demirbas, 2005). Livestock dung is a possible source of animal feed in light of the stark disparity between the availability and demand of protein feed. Before being utilized as feed, livestock feces must be harmlessly treated to remove any heavy metals and likely dangerous germs. It is clear that using feed as a means of sustainable growth and agricultural waste utilization is not a technical direction. Pigeon excrement has been used successfully as a direct feeding method by researchers. Pigs can grow and develop more quickly thanks to the abundance of proteins, amino acids, and essential mineral ingredients found in pigeon feces. Dehydrating cattle manure is the drying procedure. The moisture in the excrement was reduced until it was dry using natural, biological, mechanical, and other treatment techniques, after which it was given to animal feed. The breeding industry's waste is fermented by adding biological agents and other excipients under particular temperature and humidity conditions before being fed to poultry and fish. This technique can achieve resource recycling and is inexpensive and simple to market. It was discovered that the fermentation process may save over a third of the raw resources when compared to the standard production method. But many academics do not support this approach of using valuable agricultural trash (Demirbas, 2005).

10.8   MICROBIAL BIODEGRADATION

One of the most well-known and effective methods for the management of agricultural waste is composting. These leftovers provide a lot of benefits, including improved soil fertility and health, which boost agricultural output and soil biodiversity. Crop residues must have a C:N ratio of 30–35 and a moisture content of 55%–65% in order for microorganisms to decompose and change it into OM. These conditions are necessary for composting utilizing bacteria (Shilev et al., 2007). Many different fungal and bacterial strains that are both aerobic and anaerobic have been proven to have cellulose-related activities. A study identified several bacterial and fungi species that are involved in the breakdown of cellulose, including *Chaetomium, Myrothecium, Trichoderma, Fusarium, Aspergillus, Penicillium,* and *Trichonympha,* as well as *Clostridium, Butyrivibrio fibrisolvens,* and *Bacteroides succinogenes.* According to a different study, a microbial consortium made up of *A. nidulans, T. viride, A. awamori,* and *P. chrysosporium* was present during the composting of different crop residues with chicken manure and rock phosphate (percent), which results in nitrogen-enriched phosphocompost in 2 months (Sun et al., 2020). The bacteria that control the mesophilic stage of composting come from the Pseudomonadaceae, Enterobacteriaceae, Streptomycetaceae, and Erythrobacteraceae families. These organisms prefer a temperature range of 15°C–35°C and feed on soluble materials like sugar, amino acids, and lipids. Exothermic reactions that follow metabolic processes raise the temperature of composting to the thermophilic phase, which is 65°C–85°C. Hydrolytic enzymes are used to break down lignin, cellulose, hemicelluloses, and proteins in pseudo-nocardiaceae and thermo-monosporaceae. *A. nidulans,*

Scytalidium thermophilum, and *Humicola* sp. were found to work together in a thermophilic consortium that was particularly effective in breaking down rice straw and soybean waste. A psychrotrophic microbial consortium made up of *Eupenicillium crustaceum, Paceliomyces* sp., and others improved the composting of rice straw at low temperatures.

10.9   ECO-FRIENDLY TECHNIQUES FOR MANAGING PADDY STUBBLE

Both conventional and contemporary stubble management are components of sustainable paddy stubble management. To encourage more private players to establish facilities in their communities that might use stubble, creative ideas and actions are required. In addition, there are numerous new technologies that need to be properly assessed and adopted. Fungi are the most significant sort of group among microbiological agents and quickly colonize solid things because of this. As a result, fungi are crucial to the biodegradation of organic wastes that are derived from lignocellulosic materials. Numerous investigations have demonstrated that a number of fungal species exhibit lignocellulolytic activity (Shilev et al., 2007). Many other organisms have been investigated, including *Fusarium* sp., *A. niger, Paecilomyces fusisporous, Micromonospora,* and *Coriolus versicolor.* The lignocellulolytic microbe is viewed by some experts as a strategically significant technology for hastening the degradation of lignocellulose in agricultural wastes. In addition, it was found that a mixed culture of *R. oryzae, A. oryzae,* and *A. fumigatus* could be utilized to break down paddy straw and create superior compost with a higher concentration of macronutrients.

10.10   TRADITIONAL COMPOSTING OF AGRICULTURAL WASTE

Due to the country's rapidly expanding agriculture business, agricultural waste is growing yearly. When compared to its manufacturing pace, the recycling rate for this resourceful waste is pitiful. If agricultural waste is not properly handled, it will severely pollute the environment and pose a serious hazard to human health. On the other hand, the majority of agricultural waste is high in OM, nitrogen, phosphorous, potassium, and other essential nutrients for crops. For the creation of organic fertilizer, it is a good source. If we can collect, process, and then use the billions of tons of agricultural waste produced each year, it will be a significant source of wealth. Common methods for creating organic fertilizer from agricultural waste include composting and aerobic fermentation. Traditional composting, however, has a number of drawbacks, including a huge surface area, a protracted fermentation cycle, a significant amount of odor pollution during the fermentation process, low quality, and a small market for low-grade goods. The unfavorable reputation is limiting the process's widespread marketing and use. Utilizing this clever agricultural waste through composting is effective. Composting, in contrast to other methods of agricultural waste treatment, can not only eliminate odor, toxic materials, and pathogens bacteria, but also produce compost product that can be used as soil amendment, enhancing and improving soil structure, increasing the geochemical process of crop nutrients, and increasing soil fertility. Despite the fact that composting has several benefits for handling animal waste, traditional aerobic composting of animal waste requires a temperature of at least 50°C and a processing time of 5–10 days to ensure that the manure is sanitized safely (Shilev et al., 2007). In some cases, standard aerobic composting is not suited for breakdown because of the physiochemical features of organic waste. These qualities result in a reduced temperature increment during the composting process, which affects compost quality and cleanliness and causes insufficient decomposition. Composting has many advantages, but traditional aerobic composting can also produce a variety of undesirable materials and gases, including NH_3, H_2S. Additionally, it frequently takes a long time for organic materials to completely decompose on a vast land area, and a significant amount of organic nitrogen and carbon is lost throughout the composting process. However, by raising the treatment temperature beyond 70°C, it only takes 10–30 minutes to eliminate pathogens in animal dung. Due to the high nitrogen concentration in cattle manure, there are numerous gaseous emissions of nitrogen compounds, including ammonia (NH_3), nitrous oxide (N_2O), nitric oxide (NO), methane (CH_4), volatile organic compounds (VOCs), and others. The primary component of gaseous emissions is NH_3, and conventional composting lost a significant amount of nitrogen resources in the form of ammonia emissions, which can vary from 70% to 88% (Shilev et al., 2007). These odor emissions affect the environment since NH_3 is an unpleasant, irritating, and toxic gas. Small particle precursors are created by atmospheric NH_3 (PM2.5). Acidification and eutrophication of ecosystems are caused by NH_3 disposal. The compost's fertilizer value is further diminished by the NH_3 loss. In light of the discussion above, it is crucial to research quick and risk-free ways to create high-quality compost. A promising method for dealing with this enormous quantity of useful organic waste is ultrahigh temperature aerobic fermentation.

10.11 AEROBIC FERMENTATION PROCESS

Composting or aerobic fermentation are equivalent concepts. The aerobic fermentation process of organic waste differs from the conventional composting process in that it uses effective microbes to break down the organic waste. The aerobic composting process is actually an aerobic fermentation process because it uses a microbial agent. Depending on the type of organic wastes, the organic substrate for the fermentation process can be chosen. For the fermentation of organic waste, the powdered forms of sawdust, rice husk, coconut husk, rice bran, and other materials can be utilized as the substrate. The bacillus that is resistant to extreme temperatures, a lack of carbon dioxide, and oxygen is dried to create the long-lasting microbial fertilizer inoculum. When injected with compound bacteria during the fermentation of organic waste, the results indicated the inoculum's actions clearly, and they quickly obtained completely degraded high-quality organic fertilizer. Organic fertilizer developed by combining sawdust and chicken dung, and then adding a compound bacteria agent to carry out fermentation. Other researchers combined the powdered straw with various substances before adding some aerobic microbes to carry out aerobic fermentation and create high-quality fertilizer in a short period of time (Chen et al., 2011). When the organic waste has undergone fermentation and has a moisture level of about 30%, it can be immediately packaged as finished organic fertilizer. To improve the soil environment and make general farmland better suited for the production of "green food," the finished organic fertilizer can be applied.

10.11.1 FACTORS AFFECTING AEROBIC FERMENTATION PROCESS

Temperature—one of the most important elements in the fermentation process and the activity of microbes. Temperature has a direct impact on the maturity of the fermentation materials and the fermentation cycle by influencing the activity of microorganisms and the rate at which OM decomposes. The primary cause of the shift in temperature is the heat produced by microbial breakdown of organic materials. The temperature of the compost can be broken down into three stages during the first aerobic fermentation cycle: heating, high temperature, and cooling. Microorganisms' development states at various temperatures (Chen et al., 2011). A temperature that is too high or too low will hinder the growth of microorganisms during the entire composting process. The microorganism's enzyme activity will diminish at temperatures below 25°C. The level of decomposition will decline, lengthening the time until composting is mature. As a result, depending on the microorganisms utilized for aerobic fermentation during composting, the fermentation temperature must be managed. The pathogenic bacteria, parasite eggs, and other toxic and hazardous components in the compost can all be killed by the high temperature created during the fermentation process. To attain the harmless standard, the composting process needs to be kept at 50°C–60°C for around 5–7 days. The temperature and heat duration requirements for eradicating frequently encountered pathogens during the composting process of OM (Chen et al., 2011).

The carbon–nitrogen ratio, also known as the C/N, has a significant impact on the microbial processes involved in aerobic composting. The energy source for microbial activities is carbon. The majority of the carbon during fermentation is transformed into CO_2 and released into the atmosphere while the remaining portion is used to create the cell membrane of bacteria. The nutrition for microbial activity and a requirement for microbial reproduction is nitrogen supply. When the carbon content of the raw material for composting is too high and the nitrogen content is too low, the excess nitrogen is lost as ammonia, which has negative effects on the environment, lowers the effectiveness of nitrogen fertilizers, and lowers the quality of the finished composting products. When the C/N ratio of the composting raw material is too high, the absence of a nitrogen supply prevents microorganisms from growing, which lowers the process temperature, slows the breakdown of OM, prolongs fermentation, and lowers the nitrogen content of compost products. Later, it will lead to crops using too little nitrogen throughout the growth process and place microorganisms in the upper soil in a "nitrogen hunger" state, which directly impacts crop growth. Since the C/N ratio of the microorganism is 4–30, the raw materials' C/N ratio should likewise be kept within this range. The widespread consensus is that the initial stage of fermentation is best served by a C/N ratio of 30–35:1. The fermentation process is highly challenging since the C/N ratio of a single substance is not optimum for bacteria' regular metabolism. As a result, materials with a high carbon content or a high nitrogen content can be added to change the C/N ratio. Due to their high carbon to nitrogen ratios, straw and sawdust can be employed as high carbon additions. Contrarily, because of its high nitrogen concentration, which efficiently promotes the proliferation of aerobic microorganisms and expedites compost maturation, livestock manure can be utilized as a high nitrogen supplement. C/N can be used as a measure of compost maturity after fermentation. Several tests revealed that compost was mature when studies showed that compost could be considered mature when C/N is between 15 and 20 (Chen et al., 2011).

10.11.1.1 Ventilation

One of the main elements restricting the fermentation process and essential to the success of the aerobic fermentation of organic waste is oxygen delivery. The following succinct statement sums up ventilation's main purpose during aerobic fermentation: (i) It gives microorganisms access to enough oxygen. Consider a situation where the oxygen supply is insufficient or the contact between the fermentation materials and oxygen is uneven. The main cause of the fermentation process' unpleasant odor is that it will trigger local anaerobic fermentation and create gases like hydrogen sulfide. Additionally, the aerobic process will generate organic acids and prevent the growth of microorganisms, slowing down the breakdown of OM and lengthening the fermentation period. (ii) The temperature of the fermentation tank is likewise managed by it. By removing the heat produced by the deterioration of OM, ventilation can lower the temperature at which compost is being produced. (iii) In addition, it eliminates CO_2 and produces gases that have an impact on microbial activity while evaporating the moisture content of the composting material. The air volume should be maintained between 50 and 300 L/m^3 when the stack is being forcedly vented. At the beginning of the fermentation process, 0.2–0.5 $L/m^3 kg$ of air is ideal (Chen et al., 2011).

10.11.1.2 Moisture Content

The presence of water is a requirement for all living things. The quality of fermentation and compost products is directly impacted by changes in the materials used in fermentation. It affects the aerobic composting process physically and is a crucial aspect. The main roles of water in the fermentation process are as follows: (i) breaking down OM into small molecules and taking part in microbial metabolism activities; (ii) removing heat from composting fermentation through evaporation of water; (iii) offering a medium for biochemical reactions; and (iv) adjusting the space between materials in the composting process to balance water content and oxygen content. Throughout the entire composting process, moisture level is a critical factor. When the water content is too low (<40%), the microbial activity is inhibited, and the OM is difficult to decompose. Over 75% water content makes it difficult to raise the fermentation temperature and considerably reduces the amount of OM breakdown. In addition, too much water causes things to become soaked, which impairs ventilation and oxygen delivery, resulting in a low oxygen condition in which aerobic microbes cannot develop and produce odorous gases like H_2S. It is commonly accepted that the aerobic composting fermentation process can proceed without difficulty in the initial stage of composting when the relative moisture content is 40%–70% (by mass), and the ideal moisture content is 55%–65%.

10.11.1.3 The pH Value

It serves as the reference point for calculating the solution's acidity and alkalinity. When composting aerobically, temperature and duration both affected the compost's pH level. During the early stages of fermentation, microbial activity breaks down OM to produce organic acids, which helps to keep the compost's pH level around 6.5.

As the temperature rises, volatile organic acids will be volatilized, and microbial activity breaks down nitrogenous OM to produce NH_3, which raises the pH level of the compost to a maximum of 8.5. Appropriate pH levels can promote aerobic microorganisms' regular activity. The efficiency of the composting process and the final products as fertilizer will be impacted by an improper pH value. The pH value should be between 6.7 and 8.5 during the high-temperature portion of the fermentation process in order to achieve the greatest composting rate, according to general consensus.

10.11.1.4 Particle Size

The oxygen must travel through the space between the composting elements in aerobic composting. The amount of surface area that composting materials have in contact with microbes is indirectly indicated by particle size. The particle size requirements cannot be generalized since composting materials vary in their physical characteristics and shape. The consensus is that when the particle size of composting materials is between 12 and 60 mm, fermentation requirements can be reached.

10.11.1.5 Increasing the Performance of Bio Organisms by Adding Inert Materials

In addition to microorganisms that are safe, affordable, and well-tolerated by farmers, numerous scientists have proposed safe inert substances for handling this agricultural waste. Then, a composing technology is used, in which trash is processed to eliminate bio-toxic chemicals via anaerobic decomposition. According to earlier research, vermin-compositing is a suitable technology for turning agri-waste into a useful good. Earthworms and the accompanying microorganisms transform excrement into soil nutrients during vermin-compositing, increasing the soil's fertility (Lohan et al., 2018). Mix hazardous substances, such as pesticides, fertilizers, etc. in a 1:1 ratio with certain bulky inert ingredients, such as sawdust and cow manure, to create trails that rodents can use to compose for optimal breakdown by earthworms. Decomposition is further accelerated by combining textile sludge with biogas plant slurry at a 2:3 ratio. The performance of earthworms and the final product's nutrients are improved in a 1:1 ratio when waste from the beverage sector and cow manure are combined. Accordingly, all of these investigations showed that the performance of vermin-compositing is improved when certain bulky inert components are present in the proper ratios.

10.12 TECHNIQUES FOR BOOSTING THE SURVIVAL RATE OF MICROORGANISMS

The survival of bio-decomposers in soil affects the viability of microorganisms. Bioinoculant formulation optimizes field breakdown rate, prolongs storage time, and yields high-quality inoculants. According to US Patent No. 7097830, a bioinoculant that combines *T. harzianum* and Bacillus strains with a carrier to promote plant development. It is possible to combine several kinds of bio-inoculants in formulated products to create cost-efficient and productive crop-growing systems (Lohan et al., 2018). Additionally, this product's formulated form would keep the straightforward usability and handling of the microbial bioinoculant, which is crucial for agriculture.

For Indian farmers, waste decomposer is quite valuable. If it is transformed into formulations like wettable powder and suspension concentrate, its value will rise even more. By adding carriers, stabilizers, wetting and dispersion agents to formulated products, their viability, shelf life, and efficiency would be further improved.

10.13   STALLED PROBLEMS IN MANAGING PADDY RESIDUE

In Asia, it is common practice to burn trash outdoors, according to a research. The most typical method of burning paddy crops is in situ burning. As a result of the practice of burning rice wastes, harmful bacteria and soil invertebrates are wiped out as well as severe air pollution. Soil-borne pathogens and pests are eliminated as a result of the burning process (IP abrol). The paddy stubbles or residues aid in the adsorption of dioxins, furans, and other gaseous pollutants, as well as the grounding of greenhouse gas emissions and other gaseous pollutants including NO_3, SO_2, and HCl. As a significant generator of aerosol particles like PM10 and PM2.5, it also has an impact on the radiation budget and the quality of the local air. Farmers must quickly implement environmentally friendly rice stubble management due to the previously mentioned issues with open-field burning. Furthermore, the Rice–Wheat Consortium's experts are continuously pleading with farmers to stop burning paddy stubble (Lohan et al., 2018).

10.14   VALUABLE BYPRODUCTS FROM LARGE-SCALE AGRICULTURAL PROCESSING WASTES

Agricultural wastes can be effectively converted into manure and value-added goods rather than being burned because they don't include any non-biodegradable wastes. They can be utilized to provide green energy. In order to ensure sustainability, prevent the depletion of resources, and use solid wastes as biomass for bioenergy, a newer method of converting waste into riches has been developed. Solid state fermentation of agro-industrial wastes provides an additional choice. According to a study (Lohan et al., 2018), wheat bran, orange pulp and peel, and sugarcane bagasse were all transformed into enzymes. Increased yield, improved soil pH, nitrogen, phosphorus, potassium, and soil organic carbon levels, as well as improved crop productivity all came about as a result of mixing wet biomass and crop residue in various ratios (50:50, 50:100) in plots and different sub-plots. This method also helped to promote environmental sustainability. In a study, the bioresources from arable field crops, livestock products, and residues were predicted to be converted into usable byproducts; it was determined that rice straw and husk, rubber leftovers, and cattle dung would offer a sizable portion of Thailand's energy supply. In a study conducted in South Korea, soil enzyme activity was tested in a submerged rice cropping system due to the addition of green manure (crop residue along with amendments); increased enzyme activity was noted, along with an increase in the composition of microbial groups, microbial biomass, and enzyme activities of the soil. In a study conducted in South Africa, conventional tillage, crop rotation, and residue management affected the enzyme activity in the soil and showed that tillage and residue management were the main factors influencing soil biological indicators; there was also a significant difference in soil biological indicators under non-tillage and retention of residue. By supplying nutrients and oxygen, this method of biotillage encourages root growth. A study in South Africa, where crop rotation and residue management impacts were identified for soil quality, demonstrates, however, that crop rotation had more substantial effects on the soil quality index than residue management. The "Conservation Agriculture" technique allows for minimal soil disturbance while retaining leftovers for the soil cover. Weeds can be handled in India using traditional farming techniques such as tillage; conservation agriculture also decreases weeds via hand weeding and herbicide application. In the Shivalik highlands, plants including Guinea grass, Khuskhus, and Bhabar have been employed as "vegetation strips." They produce an upright, stiff, and dense hedge to stop overland water flow, acting as an efficient barrier against erosion and sediment control. Farmers also gain economically from them (Khuskhus). For the management of agricultural residues, additional techniques like mowing, rolling, roll-chopping, undercutting, and rototilling are used. Sometimes cover-crop residues are left on the soil's surface as a mulch; this prevents soil erosion, improves water infiltration, and controls weed growth. This is primarily done to lessen the need of chemical approaches. Another alternative for management is to produce biofuel from non-food feedstocks like agricultural wastes, forest wastes, crop residues, etc. According to a study, crop residues should be biologically pretreated and then digested anaerobically before being applied to plants. Another study examined the short- and long-term impacts of sugarcane field burning, vinasse-renewal of sugarcane field, and a combination of the two on soil OM properties. The findings demonstrated that applying vinasse helped to protect and rebuild the soil after a fire. Following the application of vinasse, the reduction in organic carbon, particle organic carbon, humic acid, and OM caused by fire was later regained. Crop leftovers from cotton ball, soybean husk, mustard husk, and straw have been reported to undergo thermochemical conversion in order to produce syngas and potassium-based fertilizer. The process of turning biomass into ethanol typically involves many steps: pre-treatment to remove lignin and hemicelluloses, enzymatic breakdown of cellulosic wastes to sugars, fermentation to produce the ethanol, and purifying of the ethanol and byproducts.

10.15 BIOCHAR OR ARBUSCULAR MYCORRHIZAL FUNGI OR COMBINATION OF BOTH AND THAT OF BACTERIA

The contribution of the soil to carbon sequestration has been investigated and published. This green restoration method, which combines biochar with AMF, has been recommended as an appropriate and long-term solution. Additionally, it has been noted that their combined application may boost biomass productivity, crop performance, soil quality, root zone system, overall surface area for the retention of water and nutrients, and decrease reliance on chemical fertilizers. In a different study, fertilizers, mustard residue, and Sesbania green manure were applied to the soil for a 5-year period. Sesbania green manure increased the soil's organic carbon content, potential for carbon sequestration, soil permeation rate, and NPK status, which was demonstrated by an increase in plant height, branches, seeds, and oil and seed yield when compared to fallow mustard practice. Another study used earthworms (*Eisenia fetida*) to vermicompost crop residues. Mustard residues, sugarcane waste, and cow dung were all treated in different lots with varying concentrations of earthworms. Additionally, the research revealed that when combined with cow manure, agricultural leftovers can become vermicompost. In accordance with a study, *Eudrilus eugeniae* can also be utilized to vermicompost crop residues such paddy straw, maize stover, leaf litter, vegetable waste, and dried flowers from temples. The mature vermicompost was microbiologically active in terms of population and enzyme activities. By adsorbing phosphorus from source-separated urine, leftover maize biomass was employed in another work to generate phosphate biofertilizer. Magnesium was added to the biochar to modify it, and the performance of the modified biochar and the unmodified biochar were compared. This was done to assess the biochar's ability to absorb phosphate. In addition to being applied to plants, biochar also had a pathogen-disinfection effectiveness of close to 80%. Crop leftovers have the added benefit of containing significant levels of cellulose and lignin, two types of biochemical energy. It might be possible to turn this biomass into biofuels. It is also possible to convert lignocellulosic biomass in a microbial fuel cell by adding rumen fluid. When used as "green manure biomass" on rice fields, leftovers from two or three different plants can enhance yield and microbial activity more than inorganic fertilizers. Microbes found in green manure can boost the soil's nitrogen, phosphorus, and carbon cycle as well as enzyme activity. Biochar is produced when fruit peels are torrefied at various temperatures for varying lengths of time. The temperature of the torrefaction process affects the properties of the biochar. By displacing coal in power plants, this can lower greenhouse gas emissions even more. Application of plant residue enhances nitrate immobilization by soil microbial organisms. In turn, this lessens nitrate buildup and nitrogen losses. After the residue of a novel grass was added, nitrate immobilization by bacteria and fungi both increased. The characteristics of crop residual waste are impacted by increased residue inputs, microbial community composition, and nitrate immobilization. Utilizing cellulolytic nitrogen-fixing bacteria for composting lignocellulosic crop residue is still in the research stages, but it has the potential to manage agricultural residue while producing high-quality compost as a byproduct. Additionally, the cellulose-degrading and nitrogen-fixing abilities of bacteria may help with quick composting, improvement of soil fertility in addition to plant growth, and long-term management of lignocellulosic crop residue. Compost can be used alone or in conjunction with bacteria, yeast, actinomycetes, oomycetes, and fungus to act as biological control agents by enhancing soil quality and plant health. The environmental damage caused by the use of nitrogen-based chemical fertilizers can be reduced by using environmentally friendly techniques, such as the nitrogen-fixing bacteria *Rhizobium*, *Pseudomonas*, *Azospirillium*, and *Bacillus*. Thus, sustainable agriculture methods are accomplished. These increase both above- and below-ground biomass.

10.16   USE OF BIO-DECOMPOSER

The National Centre for Organic Farming created a waste decomposer culture that breaks down organic waste while also enhancing soil health and functioning as a protective agent. It is a collection of microorganisms that have been taken from cow manure. Mother culture strains of various microorganisms used in the production of biofertilizers must be maintained in order to meet the standard demand. It takes between 200 and 250 different strains of microorganisms, along with regular sub-cultures in the lab. 200 L of water and 2 kg of jaggery are combined to create the beginning culture for the decomposer solution. The container is filled with one bottle of the waste decomposer (about 30 g), which is then thoroughly mixed. A wooden log should be used to stir this once or twice each day. When the solution turns creamy, which takes around 5–6 days, it is prepared for use. Spread over a plastic sheet that is positioned in the shade is 1 ton of compost. A compost layer is placed on top of the solution, which is then covered with a second layer of compost. A layer of compost is applied after the solution is sprinkled once more. For between 10 and 12 layers, do this. A moisture level of 60%–65% must be preserved. At intervals of 7 days, the compost must be turned over. It can be used after 35–40 days. This compost is a deep brown hue, malodor-free, dry, and rich in other nutrients and organic carbon. It shouldn't draw insects, flies, or other pests if it is properly prepared. Normal vermicomposting and night soil composting can produce unpleasant odors like rancid butter and ammonia, which can draw mice, flies, and other animals. However, complaints of objectionable odor and recurrent supervision issues are extremely rare or nonexistent in the case of waste decomposers.

10.17 INDUSTRIAL TREATMENT AND CULTIVATION OF EDIBLE FUNGI

Applications in the light industry include papermaking, crafts, disposable dinnerware, rayon, furfural, sugar, cellulose acetate, and xylitol, to name a few. Straw has a number of benefits over ordinary wood. Currently, a few Chinese provinces with comparatively quick development have utilized industrial straw effectively. In terms of straw utilization, Jiangsu Province leads the sector. The manufacturing of wood boards made from straw made up one-fifth of the country's total production of wood boards in 2008, while the province's industrial utilization rate of crop straw was around one-twelfth. In industrial treatment, agricultural waste is primarily burned, which produces combustion emissions similar to what occurs in the paper industry. According to the common consensus, the paper industry is a significant emitter of greenhouse gases (https://www.twosides.info/renewable energy). These emissions include GHGs including CO_2, CH_4, and N_2O. Depending on the level of development of local agriculture and animal husbandry, most farmers' primary cultivation materials include wheat, corncobs, sawdust, and other agricultural byproducts. To create edible mushrooms, the straw can be processed and used. Following the correct blending of the crop straws, regulators like gypsum are added. After the sterilizing and moisture adjustment process, they are infected with bacteria, which can then be cultivated at the right temperature and humidity. After producing edible fungus, the rods are still nutrient-rich and can be converted into cattle and poultry feed or used in production processes like fish ponds. Studies show that after being processed as a waste product of the breeding business, livestock dung can be used as a starting point for growing edible mushrooms. The output of edible fungi grown on livestock dung water was shown to be significantly higher than that of fungi produced on crop straw, resulting in higher income and lessening the load of recycling the breeding industry's waste materials. They are an essential link in the chain of businesses involved in resource recycling. More importantly, the technology of producing edible fungi with straw rod is simple and can be widely promoted, increasing national income and improving people's quality of life. In China, due to the massive output of agricultural waste, rich sources, but also has the characteristics of most of the nutrients of cultivating edible fungi.

10.18   CHALLENGES TO THE BREAKDOWN OF AGRICULTURAL WASTE

a. **Nonuniform composting:** Immature or incomplete composts have a negative impact on the soil in many ways, which in turn has an adverse impact on plant development and ecosystem function. According to a related study, applying immature compost for nitrogen fixation may either emit harmful gases or compete with oxygen in the rhizosphere, which will hamper plant growth. In addition, scientist discovered that the presence of heavy metals and ammonia in immature compost causes phytotoxicity. Growing plants and microorganisms compete for oxygen in immature compost, which produces the poisonous gases NO_2 and H_2S.

b. **Leaching:** Compost is high in organic carbon content, which is particularly advantageous to plant growth. In addition to this advantageous feature, compost leachate pollutes groundwater and is bad for the environment. The compost cannot, therefore, be kept in the same location for an extended period of time.

10.19   SUMMARY AND CONCLUSION

The residues or stubbles are the threshed, condensed portion of the rice plant that is still on the field after harvest. These unwanted leftovers need to be disposed of effectively and wisely. Finally, a number of studies have demonstrated the significant economic use of wastes as fuel, animal feed, and other raw materials. However, issues brought on by the inappropriate management of crop waste still exist. The literature review indicates that if the stubbles are not put back into the soil fields, it could lead to soil mining for primarily known nutrients, a net negative balance, and nutrient shortages in subsequent crops. As a result, this is the main reason why the agricultural system's productivity of wheat and rice has decreased. As a result, the stability and sustainability of the system depend on an efficient mechanism for disposing of the wastes of these crops.

After the crop has been harvested, several methods for treating crop wastes include burning, surface retention, integration, mulching, and domestic or commercial fuel bailing. According to this review article, the government should support such practical technology through subsidies so that more farmers can work out while farming. promoting conservation agriculture and preserving agriculture by using a variety of viable solutions to deal with paddy waste. Using microorganisms, organic and inorganic additives, and decaying paddy crop, postharvest wastes can be managed in an environmentally favorable and safe manner. Additionally, there haven't been many attempts to look at numerous obstacles to the application of this technology and other concepts. The studies indicate that environmentally responsible management of the stubbles adds a new dimension in the form of cattle bedding, packaging, fodder, packaging material, biogas, fuel, paper production, mushroom cultivation, electricity generation from bio-thermal power plants, and a resulting sustainable environment. Due to a lack of commercially feasible and approved forms of mechanization and the alternative nature of paddy trash, farmers are nevertheless compelled to burn rice straw despite becoming aware of the detrimental effects of doing so at the farm level. In many parts of the world, sustainable friendly agriculture is being practiced and adopted, which requires other in-situ measures such crop residue management with straw choppers, double disc coulters, and zero-till machines, which will lessen residue burning during the rice–wheat cycle. Therefore, encouraging organic recycling methods and taking into account farmer benefits would prevent typical industrial practices from damaging and wasting precious resources. Since burning emits carbon dioxide, sulfur dioxide, nitrous oxide, and carbon monoxide, zero-tillage and happy-seeding technology are promising innovations that offer advantages including weed control, moisture retention, and pollution prevention.

The study found that using zero-tillage technology enhances farmer benefits, livelihood, and consequently poverty reduction. According to the many studies described above, Happy Seeder Technology is a workable substitute for open-field burning and, when compared to conventionally prepared plots, saves on average about USD 23 in field preparation costs. The study makes additional assertions about the investigation and promotion of microbial biodegradation as an environmentally benign method of managing the decomposition of rice stubble. Several studies were conducted to look into the current problems with microbial biodegradation of rice stubble. The review study highlighted the value of in-situ paddy stubble breakdown and the advantages of implementing such environmentally friendly rice management as a result (paddy).

REFERENCES

Abbasi, M.K.; Tahir, M.M.; Sabir, N.; Khurshid, M. (2015). Impact of the addition of different plant residues on nitrogen mineralization-immobilization turnover and carbon content of a soil incubated under laboratoryconditions. *Solid Earth*, 6, 197–205.

Awais, M.; Li, W.; Arshad, A.; Haydar, Z.; Yaqoob, N.; Hussain, S. (2018). Evaluating removal

Awais, M.; Li, W.; Cheema, M.J.; Hussain, S.; AlGarni, T.S.; Liu, C.; Ali, A. (2021). Remotely sensed identification of canopy characteristics using UAV-based imagery under unstable environmental conditions. *Environ Technol Innov*, 22, 101465.

Bardgett, R.D.; van der Putten, W.H. (2014). Belowground biodiversity and ecosystem functioning. *Nature*, 515, 505–511.

Benbi, D.K.; Nayyar, V.K.; Brar, J.S. (2006). The green revolution in Punjab: Impact on soil health. *Indian J Fertil*, 2(4), 57–66.

Brankatschk, G.; Finkbeiner, M. (2017). Crop rotations and crop residues are relevant parameters for agricultural carbon footprints. *Agron Sustain Dev*, 37, 58.

Bridgham, S.D.; Ye, R. Organic matter mineralization and decomposition. In *SSSA Book Series*; DeLaune, R.D., Reddy, K.R., Richardson, C.J., Megonigal, J.P., Eds.; American Society of Agronomy and Soil Science Society of America: Madison, WI, 2015; pp. 385–406.

Chen, B.; Liu, E.; Tian, Q.; Yan, C.; Zhang, Y. (2014). Soil nitrogen dynamics and crop residues. A review. *Agron Sustain Dev*, 34, 429–442.

Chen, T.; Wu, C.; Liu, R.; Fei, W.; Liu, S. (2011). Effect of hot vapour filtration on the characterization of bio-oil from rice husks with fast pyrolysis in a fluidized-bed reactor. *Bioresour Technol*, 102(10), 6178–6185.

Ciais, P.; Sabine, C.; Bala, G.; Bopp, L.; Brovkin, V.; Canadell, J.; Chhabra, A.; DeFries, R.; Galloway, J.; Heimann, M.; Jones, C.; Le Quéré, C.; Myneni, R.B.; Piao, S.; Thornton, P. (2013). Carbon and other biogeochemical cycles supplementary material. In *Climate Change 2013: The Physical Science Basis. Contribution of Working Group I to the Fifth Assessment Report of the Intergovernmental Panel on Climate Change*; Stocker, T.F.; Qin, D.; Plattner, G.K.; Tignor, M.; Allen, S.K.; Boschung, J.; Nauels, A.; Xia, Y.; Bex, V.; Midgley, P.M., Eds. https://www. climatechange2013.org and www.ipcc.ch.

Demirbas, A. (2005). Potential applications of renewable energy sources, biomass combustion problems in boiler power systems and combustion related environmental issues. *Prog Energy Combust Sci*, 31(2), 171–192.

Galus-Barchan, A.; Chmiel, M.J. (2019). The role of microorganisms in acquisition of nutrients by plants. *Kosmos*, 68, 107–114.

Hegde, N.G. (2010). Forage resource development in India. In: *Souvenir of IGFRI Foundation Day*, November, 2010. www.baif.org.in.

Henriksen, T.; Breland, T. (2002). Carbon mineralization, fungal and bacterial growth, and enzyme activities as affected by contact between crop residues and soil. *Biol Fertil Soils*, 35, 41–48.

Jamir, E.; Kangabam, R.D.; Borah, K.; Tamuly, A.; DekaBoruah, H.P.; Silla, Y. (2019). Role of Soil microbiome and enzyme activities in plant growth nutrition and ecological restoration of soil health. In *Microbes and Enzymes in Soil Health and Bioremediation*; Kumar, A., Sharma, S., Eds.; Microorganisms for Sustainability; Springer: Singapore, pp. 99–132.

Jin, Z.; et al. (2020). Effect of straw returning on soil organic carbon in rice-wheat rotation system: A review. *Food Energy Secur*, 9(2), e200.

Jones, E.; Jessop, R.S.; Sindel, B.M.; Hoult, A. (1999). Utilising crop residues to control weeds. In *Proceedings of the 12th Australian Weeds Conference, Papers and Proceedings, Weed Management into the 21st Century: Do We Know Where We're Going?* Hobart, Tasmania, Australia, 12–16 September 1999; University of Tasmania: Hobart, Australia.

Khan, A.; et al. (2009). Biomass combustion in fluidized bed boilers: Potential problems and remedies. *Fuel Process Technol*, 90, 21–50.

Lohan, S.K.; Jat, H.S.; Yadav, A.K.; Sidhu, H.S.; Jat, M.L.; Choudhary, M.; Sharma, P.C. (2018). Burning issues of paddy residue management in north-west states of India. *Renew Sustain Energy Rev*, 81, 693.

McMurtrie, R.E.; Iversen, C.M.; Dewar, R.C.; Medlyn, B.E.; Näsholm, T.; Pepper, D.A.; Norby, R.J. (2012). Plant root distributions and nitrogen uptake predicted by a hypothesis of optimal root foraging: Optimal root foraging. *Ecol Evol*, 2, 1235–1250.

Pandiaraj, T.; Selvaraj, S.; Ramu, N. (2015). Effects of crop residue management and nitrogen fertilizer on soil nitrogen and carbon content and productivity of wheat (*Triticum aestivum* l.) in two cropping systems. *J Agric Sci Technol*, 17, 249–260.

Pangnakorn, U.; George, D.L.; Tullberg, J.N.; Gupta, M.L. (2003). Effect of tillage and tra_c on earthworm populations in a vertosol in south-east Queensland. In *Proceedings of the Soil Management for Sustainability*, Brisbane, Australia, 13–18 July 2003; International Soil Tillage Research Organisation: Queensland, Australia.

Paolini, V.; Petracchini, F.; Segreto, M.; Tomassetti, L.; Naja, N.; Cecinato, A. (2018). Environmental impact of biogas: A short review of current knowledge. *J Environ Sci Health A*, 53(10), 899–906.

Paul, E.A. *Soil Microbiology, Ecology and Biochemistry*, 4th ed.; Academic Press: Waltham, MA, 2014.

Paul, K.I.; Polglase, P.J.; O'Connell, A.M.; Carlyle, J.C.; Smethurst, P.J.; Khanna, P.K. (2003). Defining the relation between soil water content and net nitrogen mineralization. *Eur J Soil Sci*, 54, 39–48.

Purohit, P.; Chaturvedi, V. (2018). Biomass pellets for power generation in India: A techno-economic evaluation. *Environ Sci Pollut Res*, 25(29), 29614–29632.

Rashid, M.I.; Mujawar, L.H.; Shahzad, T.; Almeelbi, T.; Ismail, I.M.I.; Oves, M. (2016). Bacteria and fungi can contribute to nutrients bioavailability and aggregate formation in degraded soils. *Microbiol Res*, 183, 26–41.

Shilev, S.; Naydenov, M.; Vancheva, V.; Aladjadjiyan, A. (2007). *Utilization of By-products and Treatment of Waste in the Food Industry*, Springer: Boston, MA.

Singh, R.P.; Dhaliwal, H.S.; Humphreys, E.; Sidhu, H.S.; Singh, M.; Singh, Y.; John, B. (2008). *Economic Assessment of the Happy Seeder for Rice-Wheat Systems in Punjab, India*. A paper presented at AARES 52nd annual conference, Canberra, ACT, Australia.

Strickland, M.S.; Lauber, C.; Fierer, N.; Bradford, M.A. (2009). Testing the functional significance of microbial community composition. *Ecology*, 90, 441.

Sun, M.; Xu, X.; Wang, C.; Bai, Y.; Fu, C.; Zhang, L.; Fu, R.; Wang, Y. (2020). Environmental burdens of the comprehensive utilization of straw: Wheat straw utilization from a life-cycle perspective. *J Clean Prod*, 259, 120702.

Surekha, K.; Kumari, A.P.P.; Reddy, M.N.; Satyanarayana, K.; Cruz, S. (2003). Crop residue management to sustain soil fertility and irrigated rice yields. *Nutr Cycl Agroecosyst*, 67, 145–154.

Thakur, T. C. (2003). Crop residue as animal feed: Addressing resource conservation issues in rice-wheat systems of South Asia, a resource book. Rice Wheat Consortium for Indo-Gangetic Plains (CIMMYT), March, 2003.

Wang, B.; Shen, X.; Chen, S.; Bai, Y.; Yang, G.; Zhu, J.; Shu, J.; Xue, Z. (2018). Distribution characteristics, resource utilization and popularizing demonstration of crop straw in southwest China: A comprehensive evaluation. *Ecol Indic*, 93, 998–1004.

Wang, D.; Rosen, C.; Kinkel, L.; Cao, A.; Tharayil, N.; Gerik, J. (2009). Production of methyl sulfide and dimethyl disulfide from soil-incorporated plant materials and implications for controlling soilborne pathogens. *Plant Soil*, 324, 185–197.

Yadvinder-Singh, B.S.; Timsina, J. (2005). Research Publications Repository-Crop residue management for nutrient cycling and improving soil productivity in rice-based cropping systems in the tropics. *Adv Agron*, 85, 269–407.

Index

Printed in the United States
by Baker & Taylor Publisher Services